The Challenge of S

3

The Challenge of Smallholding

SEDLEY SWEENY

Oxford   New York

OXFORD UNIVERSITY PRESS

1985

Oxford University Press, Walton Street, Oxford OX2 6DP

London New York Toronto

Delhi Bombay Calcutta Madras Karachi
Kuala Lumpur Singapore Hong Kong Tokyo
Nairobi Dar es Salaam Cape Town
Melbourne Auckland

and associated companies in
Beirut Berlin Ibadan Mexico City Nicosia

Oxford is a trade mark of Oxford University Press

First published 1985 as an Oxford University Press paperback

British Library Cataloguing in Publication Data

Sween, Sedley
The challenge of smallholding.
1. Agriculture—Great Britain
2. Farms, small—Great Britain
I. Title
630   S513.2
ISBN 0-19-286018-6

Library of Congress Cataloging in Publication Data

Sweeny, Sedley
The challenge of smallholding.
(Oxford paperbacks)
Bibliography: p.
Includes index.
1. Agriculture—Great Britain—Handbooks, manuals, etc.
2. Farms, Small—Great Britain—Handbooks, manuals, etc.
3. Organic farming—Great Britain—Handbooks, manuals, etc.
4. Agriculture—Handbooks, manuals, etc.
5. Farms, Small—Handbooks, Manuals, etc.
6. Organic farming—Handbooks, manuals, etc.
I. Title.
S513.2.S87   1985   630   84-11248
ISBN 0-19-286018-6 (pbk.)

Set by Burgess & Son, Abingdon, Limited
Printed in Great Britain by
Richard Clay (The Chaucer Press) Ltd.
Bungay, Suffolk

Contents

Figures

Tables

Author's note on weights and measures

Britain's transition from imperial to metric measurements has not made it easy to adopt a consistent policy in this book. I have tried to be as practical as possible, which has meant reflecting many of the existing anomalies. Land is still sold in acres, and honey in pounds, whereas animal feed is measured in kilograms, and temperature in centigrade. Even more confusing is the fact that the Milk Marketing Board buys milk from the farmer by the litre and sells it to the consumer by the pint. You may find wood, wire, and string measured either by the metre or by the foot, depending where you buy it: I have assumed that the reader will be more familiar with feet and inches.

In order to help the reader through this maze, conversion tables are given below.

Imperial		Metric
1 inch	=	25.4 millimetres
1 foot	=	0.3048 metre
1 yard	=	0.9144 metre
1 mile	=	1.609 kilometres
1 acre	=	0.405 hectare
1 pint	=	0.568 litre
1 gallon	=	4.546 litres
1 ounce	=	28.35 grams
1 pound	=	0.4536 kilogram
1 hundredweight	=	50.80 kilograms

Metric		Imperial
1 millimetre	=	0.039 inch
1 centimetre	=	0.394 inch
1 metre	=	1.094 yards
1 kilometre	=	0.6214 mile
1 hectare	=	2.471 acres

1 litre	=	1.76 pints
1 gram	=	0.0353 ounce
1 kilogram	=	2.205 pounds

Fahrenheit: water boils at 212° and freezes at 32°
Centigrade: water boils at 100° and freezes at 0°
$F = 9C/5 + 32$
$C = 5(F - 32)/9$

Ill fares the land, to hastening ills a prey,
Where wealth accumulates, and men decay:
Princes and lords may flourish, or may fade;
A breath can make them, as a breath has made;
But a bold peasantry, their country's pride,
When once destroyed, can never be supplied.

Oliver Goldsmith, *The Deserted Village*

1 Introduction

A nation's survival as a free, independent, and self-respecting entity hangs on the ability of its people to nourish and protect themselves; to provide the means of building and maintaining healthy minds in healthy bodies, and to develop the enterprise, resilience, and determination to surmount natural disasters and adapt to ever-changing conditions. The spirit and morale of a nation has no foundation more important than the self-reliance and confidence of its individual people. How can we possibly retain that self-reliance and confidence if we become more and more a nation of programmed consumers, stuffed with the produce of an automated technology over which we feel we can have little influence? And what assurance can there be that the brittle, complex industrial structure will not collapse through the failure of some small but vital component? Indeed many fear that collapse is happening already.

How can this frightening slide towards hopelessness be reversed? How can we regain individual and group self-reliance?

The first requirement must be love of and pride in one's country, which can grow only in men and women who genuinely feel they have a stake in it, however small; men and women who know that their own initiative, ingenuity, and hard work will improve that stake and make fuller and more secure lives for their families.

To provide people with the opportunity to acquire even a tiny stake in the land is a daunting task calling for wisdom, courage, and political integrity almost unheard of in modern times. Indeed, to do so would mean the reversal of a process that has, since the sixteenth century, relentlessly parted the commoner from his freehold and common rights, and reduced him to a landless labourer. Today, with less than $2\frac{1}{2}$ per cent of Britain's people left on the land (and this percentage is still falling), the destruction of the 'bold peasantry' might be considered complete; that it is too late to revive and rebuild it. But this is a counsel of despair; an admission that our country is past praying for.

Fortunately the dogged good sense of the ordinary man is far from dead. The instinctive urge to acquire a stake in the country, to get back to first principles and become, once again, master of his own fate, becomes ever more apparent as our top-heavy, complex society lurches from crisis to crisis. A small but growing number of sane, apparently successful men and women are voluntarily abandoning the struggle for material gain and accepting the stern discipline of relearning the near-forgotten skills of smallholding.

Since the turn of the century there have been brief periods when Britain has been more or less self-sufficient in cereals, meat, and dairy produce – during the two World Wars, and today, under the extraordinary economics of the European Community. In each case the land has been exploited far beyond its true potential – out of dire necessity during the Wars, and now for blind greed and short-term gain resulting in vast surpluses which cannot be eaten or sold. As A. G. Street so clearly demonstrated in his classic *Farmer's Glory* in the 1930s, these indulgences have invariably been followed, quite suddenly, by much longer periods of agricultural depression when the land has had to be grassed down to rebuild fertility. There can be little doubt that this is about to happen again, and our debt to the soil is far greater than ever before. It has long been known that skilled smallholders, owning their own plots of land, can produce very much more from an acre than large-scale farmers with labourers who cannot possibly have the same dedication. What is more, a good smallholder, in his own interest, will develop a system of husbandry that can be sustained indefinitely without robbing the soil. How else could he retain his freehold to pass on to his descendants? The Chinese have maintained a stable agriculture for 4,000 years and feed more than four times as many people per acre as we do. Their system is based on a vast number of smallholders on tiny plots of land, working with incredible skill, dedication, and thrift.

What can be done to give even a tenth of our people the opportunity to acquire a small stake in the land, and to replace the missing bottom rungs in the farming ladder?

The present climate is not promising. The cost of freehold land is so high that only a wealthy man who farms intensively can hope to make a living on his own small farm. The increasing acquisition of land by large institutions reduces the number of holdings. The tax and succession laws make it virtually impossible for landlords to create smallholdings on their estates, and the Land Tenure Acts (designed to protect tenants) have dried up new tenancies. Many County Councils are selling off their statutory smallholdings; on the rest the rents are so high that the tenants are forced into intensive husbandry. Government and EEC subsidies result in gross over-capitalization on nearly every farm in the land. So, in the short term I see little hope for the newcomer or success for the organic smallholder. In the longer term I am more optimistic. The present idiotic trends cannot go on forever, and there could soon be drastic changes. Can these changes stem from enlightened rethinking of principles or must they follow a collapse of the EEC and the present economy?

There *are* steps (not politically attractive) that our Government can take to bring about successful changes. First, they must *want* to see a larger number of small farms (at present they don't) and this means discouraging land management by large institutions. Second, the tax and succession duty laws need amending to enable landlords to create smallholdings on their estates without incurring unacceptable penalties (many would welcome the chance to do so). Third, the Land Tenure Acts should be amended to make tenancies attractive to landlords. Fourth, County Councils should be forced to meet their obligations to provide smallholdings and allotments, and should not be allowed to sell them off. Finally (and this will raise a storm), a law should be enacted limiting the acreage which any individual (or institution) may farm 'in hand'.

Opposition from vested interests will be intense, but no one should underestimate the power of the goodwill that survives to be tapped as soon as politicians are seen to be moving in the right direction.

The choice is there: I believe it is between hope and eventual disaster.

The philosophy of smallholding is the philosophy of self-reliance, freedom, and independence. But this in no way implies *selfish* sufficiency, licence, or isolation. No man is an island; no smallholder can possibly find the time or produce all the skills needed for complete self-sufficiency. Nor is it suggested that the smallholder should revert to primitive subsistence living, although he may need to give up some of the consumer society's more expensive luxuries.

Co-operation with neighbours is the key to successful smallholding. It is a matter of scale; of working with like-minded, self-employed farmers and craftsmen rather than of subservience to faceless officials and giant firms. Given such co-operation, the main constituents of a full, rewarding life are available within the community, and cash farming can begin to take second place to the production of local needs. This change of emphasis from cash to produce is probably the most difficult mental hurdle. Immediate cash profit has far too long been the overriding aim: it has warped the genuine economy and forced farmers to consider every move in the context of 'How soon will it pay?' Cash profits are necessary; but the nearer a community can come to providing its own basic requirements of food, shelter, and energy, the less cash is needed for the luxuries. Within the pattern of large-scale commercial farming, the independent smallholder can still provide *more than his share* of surplus food for the urban population, yet keep as top priority the conservation of his main capital – the soil.

The acceptance of discipline is probably the greatest step towards freedom a man can take, and nowhere is this so true as in smallholding. How easy it is to let the seven-days-a-week routine become drudgery, be overwhelmed by work unfinished, and to lose sight of the original aim. This is the prelude to disaster in which soil, crops, and (particularly) livestock can suffer as much as the farmer. The failure of projects leads to parsimony which goes hand in hand with waste. The holding becomes a slave-driving tyrant.

An old naval officer once said 'True discipline implies a disciple – the happy follower of a teacher or ideal.' True discipline must be self-discipline, freely accepted. In my opinion the hardest task on the holding is to overcome inertia, to make a start on the daunting task. Once this initial discipline is faced, the job itself is relatively easy.

The habit of frugality, the avoidance of waste at every turn, is an essential part of smallholding, where margins are so narrow. This habit is more natural to some than to others, but conscious discipline is needed to extend it into every aspect of the farm. If parsimony stems from waste, so generosity can flow from a frugal husbandman.

Smallholding can be a very healthy occupation. It needs to be, for an unfit smallholder has little hope of success, and too few people know what it feels like to be really fit. It is a great blessing to possess a sound mind in a healthy body, and it is our duty to build this gift into true fitness. Strength, agility, vigour, and endurance do not just happen, but require proper nutrition and training to develop their full potential. Here, once again, discipline is vitally important. Hand in hand with physical fitness goes the courage, enthusiasm, and mental power to stick to our aims when the going is tough and soft options beckon.

Each of the many skills of smallholding requires its own discipline. 'If a job is worth doing, it is worth doing well' – so runs the adage. On a smallholding there is no room for shoddy work.

As our eyes open to the perilous state of the world, it is hardly surprising that so many sensible people are turning instinctively to the land and looking for smallholdings. But inflated land prices, which have been rising much faster than the cost of living index, have placed even smallholdings out of reach except for the well-endowed. Even well-established commercial farmers are finding themselves on a cruel tread-mill of soaring capital investment, the financing of which can only be serviced by increasing mortgages based on the rising cost of land. While this inflation lasts, it is possible for these agribusinessmen to live superficially well if their nerves can stand the strain, and their real capital, the soil, survive such prodigal exploitation. To go on thus is like walking up an ever-narrowing blind alley: the further you go the more difficult it is to turn about; the more certain and greater the ultimate disaster. A great proportion of our farmers, large and small, know in their hearts that they are on the wrong road, but have neither the wisdom nor the courage to strike out on radically new lines. To do so they would have to buck the very system on which they now depend. For them freedom and self-reliance seem very far away.

I do not believe that the answer lies in amalgamations, technology, and further reduction of manpower. The creation of paper 'wealth' through such manipulation of our economy can only spell moral bankruptcy for our people. The sane alternative must be to create conditions in which small, self-reliant enterprises can take root and grow; to provide training and guidance for those bold enough to lead in this direction. New definitions and values are needed for 'wealth', 'profit', and 'economics'. Johnson defined the latter thus: 'Economy (gr. oikos nomos); The management of a family; the government of a household.' I should like to add the rider, 'for the joy and well-being of those who dwell therein'.

The organic approach

The purpose of this book is to provide a comprehensive guide for smallholders; for beginners and those with some experience, as well as for those who are just thinking about it. The widely differing soils and climate within our countryside together with the variation in personal aspirations and abilities produce a vast range of enterprises to be considered. Is the first aim to make money (we must obviously avoid losing it!) or to provide for our personal needs? How self-sufficient is it sensible to be? Do we plan to be full-time or merely spare-time smallholders? These are but a few of the variables.

No one farmer could possibly be qualified to advise on all the options, nor could he do more than skate over them in one volume. I have therefore based this book on my own limited experience, adding, with due acknowledgement, short outlines on systems or aspects beyond my personal competence, gleaned from experts in their own particular fields.

Most farmers have their own strongly held views or even prejudices, and I am no exception. Basically the book is an account of one man's experience, his strength and his weakness, his success and his mistakes. It also points to alternatives for those who do not wholly share his views. I have been an organic farmer for twenty-five years, and look upon nature as an integrated and living entity based on an amazingly intricate and balanced plan. Darwin's concept of evolution may hold water as far as it goes, but without the inclusion of an omnipotent 'Master Mind', working from outside to plan the possibility of purposeful development, it makes no sense to me at all. I therefore decline to don the straitjacket of those scientists (not all) who can only accept such facts as they reach via their own limited, analytic logic; who tend to study the parts in ever narrowing detail but too seldom consider the whole. Farming, especially mixed smallholding, needs to be viewed in its interlocking whole. But no man can hope to attain scientific mastery of its many parts and re-assemble them into the wonderful symbiosis of nature. He must be prepared to accept a great deal on faith, glad to admit 'I don't know how, but it certainly works'. I am reminded of the ant who asked the centipede in what order he put his feet to the ground. The poor centipede worried over the answer until he had a nervous breakdown and lost the ability to walk naturally. I have no doubt that every mystery of nature has a scientific explanation that may, one day, be expressed in terms of intellectual logic. In the mean time the practical farmer must get on with his husbandry, relying on his limited knowledge and a great deal of intuitive instinct. That this intuition can be trained and sharpened I have not the slightest doubt, and I further believe that this is most likely to occur when one looks upon natural phenomena with reverence and humility.

Some years ago I discussed my beliefs with a learned Professor of Agriculture at a leading university. 'You are quite right,' he said, 'but ten years ahead of your time. If I tried to teach that here, I should get the sack.' Time is now catching up; attitudes are changing and conventional dogma is called in question. Scientific logic has put men on the moon, but seems incapable of solving the energy crisis or of providing the means whereby mankind can live purposefully on a finite planet with shrinking resources.

Perhaps the time has come when we should humbly admit our limitations and seek solutions in co-operation with the Master Plan.

2 The land of Britain

How farming developed

Once upon a time man lived by hunting and by gathering. Like other wild creatures, he moved with the seasons in search of food and made little impression on the countryside. His keen instinct gave him a deep understanding of his environment and his innate intelligence led him to devise ways of altering it to his advantage. Two main streams developed; the hunter became herdsman and the gatherer became agriculturalist. But the two life-styles were incompatible; the herdsman followed – and later drove – his animals from grazing to grazing and became a nomad, whilst the gatherer settled down to till the soil and defend his crops from predators. The integration of these systems followed gradually, and much later, when man learned to grow and store crops for his now domesticated animals that they might survive winter or drought at home. To this day the two life-styles survive separately over great areas of the Middle East and central Asia. To a lesser degree they still exist nearer home; in the Alps herdsmen take their cattle to the 'Alm' meadows where they tend them during the summer, and in Scotland and Wales the hill sheep spend half the year on the mountain commons whilst the lower, enclosed land grows winter keep. In many developing countries 'agriculture' and 'farming' are still considered to be separate subjects; the tilling of the soil and animal management have not been integrated.

Until the Industrial Revolution, farming in Britain (in its

combined sense) remained predominantly small-scale, mixed, and locally self-sufficient. Most holdings grew their own wheat and every village boasted a mill. Butter and cheese were made on practically every farm and even the cottager killed and salted his own bacon. Surplus cattle were slaughtered in the autumn to enable the breeding stock to survive on the scarce winter rations.

But all this was to change as, from the sixteenth century onwards, enclosures were enforced and estates enlarged. Wool, beef, dairy products, and cereals became more and more commercial sources of wealth, as opposed to providing for local self-sufficiency, and were concentrated in the areas most suited to their cheap production. Land drainage and the development of transport and farm machinery hastened this process, and it became 'uneconomic' for every small farm or village to aspire to self-sufficiency. Not only did farms become larger and more specialized, but village craftsmen – millers, blacksmiths, carpenters, wheelwrights, bakers, and cobblers – were squeezed out as their trades were concentrated in the towns.

Today we are in the vulnerable position of producing barely half our own food, with only 2½ per cent of our people directly concerned with working the land, and that percentage is still falling. 'Economic' considerations have forced our farming into extreme specialization where the basic rules of conservation and good husbandry (indeed, some would say ethics) are too often flouted. But if only 2½ per cent are on the land, they are supporting (and rely on) a vast population of industrial and commercial workers – over whom they have no control – who supply the machinery, power, feeding-stuffs, and fertilizers that make the specialization possible. Thus we have arrived at today's rural scene, which, with modern technology, is changing more rapidly than ever.

The British landscape

Four factors combine to make the British countryside what it is today: the geology of the earth's crust, the climatic history of the glacial and post-glacial ages, the present-day climate (particularly rainfall), and the influence of man.

The main shape of the landscape – mountains, rolling hills, and flat plains – is based on the geological rock formations, but has been severely modified by the movement of glaciers which carried vast quantities of rock, grinding it down in the process, depositing it in the form of boulders, gravel, sand, and silt, often many miles from its original outcrop. This residue has been further moved by subsequent erosion and climate, and forms the soil we know today, which does not necessarily reflect the local underlying rock formation.

The modern climate, particularly the distribution of rainfall, has combined with different soils to produce the variable vegetation which now covers our landscape. In this evolutionary process, man has played an increasing role; clearing forests, draining swamps, introducing new crops and animals, and, not least, altering the ecological balance between minerals, flora, and fauna in the soil itself.

In very general terms the western half of Britain is much higher, wetter, and more acid than the eastern half and the northern half colder than the South (see Figure 1). The strong sand and clay loams and alluvial silts of the East are more fertile than the thinner, stonier soils on the western uplands. The richer, drier East has a much longer growing season than the poorer West. Within this general pattern there are local variations; the west coast of Cumbria and Scotland, for example, is particularly affected by the warm Gulf Stream and thus has mild, wet winters, and, within the hills of the North and West, are pockets of strong land and areas of much lower rainfall. By and large, however, the best arable crops are grown in the South and East whilst the North and West are devoted to livestock rearing.

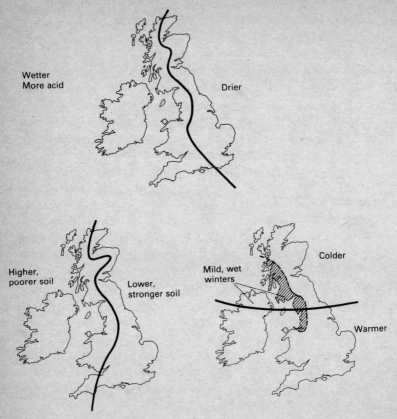

Fig. 1.   The British landscape

Hill farming

The rugged mountains and moors of Scotland, Wales, and the West Country are too steep and wet to grow good arable crops, too poor for dairying, and the grazing season is too short for fattening beasts. These areas are predominantly used to rear hardy beef cattle and sheep which are sold in the autumn to be finished elsewhere. The hill farms generally have small fields near the valley bottoms, larger rough grazing 'in-bye' areas which adjoin the open hill, and common hill land where sheep, ponies, and occasionally cattle are turned out during the summer. Grass is the main crop, supplemented by the odd field of oats, barley, or roots for the winter feeding of breeding stock. A few of the earlier lambs may be sold fat off their mothers, but the main sources of income are from six- to eight-month-old beef calves, store lambs (about six months old), four-year-old ewes which go down-country for cross-breeding, and the wool clip. There is seldom enough winter keep for the whole of the breeding stock, and ewe lambs are often sent away to winter in less rugged conditions. Hill farmers are primarily stockmen who delight more in animals than in machinery, buildings, or cash crops. Their main preoccupation is providing winter keep for their breeding stock from steep, wet, acid soils, which are expensive and dangerous to work with modern machinery. Small wonder, then, that they confine their field-work to spreading lime and slag, making hay, and reseeding the odd pasture via a short rotation of feed oats, rape, and turnips. For gathering, shearing, dipping, and sometimes haymaking, hill farmers still work a system of 'help your neighbour' in which several come together on predetermined dates to do these important jobs quickly and efficiently. Many hill farmers are smallholders with part-time jobs elsewhere.

Hill farming demands very hard work for marginal returns and were it not for government subsidies for hill sheep and cattle, many hill farmers would have to give up. As it is, the extra cost of fencing and improving steep land and buying supplementary winter feed is seldom balanced by the subsidies.

Fattening pastures

The cheapest way to fatten beef is on grass, and for this rather special conditions are required. First, we need a long, warm, unbroken growing season to provide a plentiful diet of lush herbage. Second, there must be a constant supply of water to the sward even during long spells of dry weather. Third, the ground must be well drained to prevent it being poached by the animals. These conditions are best met in low-lying areas that were once marshland, and which still lie above a plentiful water table. The most famous of these fattening pastures are in Northamptonshire, Leicestershire, and Rutland; others are in North Somerset, parts of Norfolk, the Severn Valley, Northumberland, and eastern Scotland. Many of these areas lie too wet in winter for arable farming, and thus lend themselves to bought-in herds of cattle for summer fattening. Smaller areas of good fattening pasture are to be found in other parts of the country, but before attempting to finish cattle on grass, the smallholder should ensure that the three main conditions – a long growing season, constant subsoil water supply, and good drainage – are met. A long growing season is unlikely to occur in a high rainfall area or on ground above 600 feet. A constant water supply cannot be guaranteed in sandy or rocky soils in a low-rainfall area.

Fattening pastures, as opposed to land where growing youngstock is reared or dairy cows kept, require little manure apart from the droppings of the grazing animals, as they are not exporting bone (phosphorus) or milk (calcium) from the land.

The grazier's main problems are to provide leys of palatable and nutritious grasses and clovers and to see that these are consumed in the lush, leafy state before they grow to stem and seed. As the three important conditions (see above) vary, he must be ready to buy extra stock, top pastures with a mower to prevent the grass from growing to stem and seed and to encourage the nutritious leafy portion, make hay, or on occasion buy in oil cake or other succulent feed to keep his animals going steadily ahead.

The chalk downs

The light, well-drained chalk downs of Wiltshire, the Cots-wolds, Yorkshire, and Lincolnshire were the first areas of the country to be farmed. The soil could be worked by light, primitive ploughs and there were no forests to clear or swamps to drain. But these are thin soils with an uncertain water supply; not naturally the best land for arable cash crops or cattle. Later, in the middle ages, these downs were mainly tumble-down grassland with extensive flocks of sheep. Later still, cereal and root crops were grown and the sheep were folded intensively over the roots and stubbles. Now, with the advent of sophisticated machinery and chemical fertilizers, the downs are supporting large dairy herds in conjunction with the growing of barley and wheat for sale. Water, particularly for cattle, is still a limiting factor. Productivity has been maintained, as far as possible, by the increasing use of artificial fertilizers, but it has proved very difficult to keep up the humus level on which a good soil structure depends. Straw-burning after harvest, now widely practised, exacerbates this problem, and some downland farmers are finding it necessary to go back to grass and more livestock to maintain the humus level and soil structure.

The flinty soils overlying chalk are ideal grazing land so long as water is available. They do not poach in winter, but are liable to become desiccated in summer. They will grow good cereal crops so long as soil structure is maintained by the return of sufficient organic matter.

The arable lands

The best plough lands are mainly in the broad, flat eastern half of the country from the Moray Firth to East Anglia and West Sussex. Parts of the Midlands, East Lancashire, Notting-hamshire, Shropshire, and North Cheshire also grow excellent arable crops. This land is mainly low-lying, of strong soil, low rainfall, and has a long, warm growing season. The types of soil vary from heavy clay loams to deep, fine silts (the Wash), sands, and peats (the Fens). In the well-drained driest parts, the heavy clay loams make the best arable land; in the wetter parts these clays are better kept in grass, and the sandy loams best for ploughing.

Wheat, originally a plant of the hot, arid, Near East, has been somewhat improved and adapted to temperate climates, but still needs strong land and a long, hot, dry growing season to develop and ripen. Barley thrives in somewhat cooler climes and, being less 'hungry', does not need such rich soil – it tends to grow too long and 'lodge' on very strong soils – but it still needs a longish season to ripen well. Oats and rye are much more tolerant of cold and wet and are more suitable for the poorer soils of the North and West.

As well as cereals, root crops – particularly potatoes and sugar beet – are widely grown on the best arable land, whilst in areas nearer the big cities, vegetables are most profitable, particularly on well-drained sandy loams.

The dry eastern half of Britain is generally susceptible to significant periods of summer drought, when growth ceases in shallow-rooting plants such as grasses and clovers. This makes the area unsuitable for dairy farming where a constant supply of lush summer grass is essential. Cereals and sugar beet, on the other hand, ripen well in such conditions.

Dairy farming

Ideal dairy land has a long growing season, no summer drought period, a stable soil structure with good drainage to avoid excessive poaching, and an unfailing water supply. The larger dairy herds are thus to be found in the intermediate areas between the dry, hot plough lands of the East and South and the cold, wet uplands of the North and West. Somerset, Dorset, the Severn Valley, parts of Devon, the west Midlands and the warm west coasts of southern Scotland and Cumbria are amongst the many good dairying districts. Smaller herds can be managed commercially under less ideal conditions; the smallholder's house cow, receiving individual attention, can do well almost anywhere.

Conclusion

This general description of the countryside oversimplifies what is really a very complicated pattern of soils and climate. Within the cold, wet hills of Wales and Scotland, there are pockets of much stronger land with warmer, drier climates. Next to the thin chalk downs in Wiltshire lies the Vale of Pewsey with its superb deep greensand. As a result of glacial history, wind-blown erosion of sand dunes, and silting of the river estuary, the south bank of the Ribble between Southport and Preston is one of the finest vegetable-growing areas in the country. Within the overall pattern we must look very carefully for the local variations, realize the implications of both soil and climate, and plan our holding to make the most of the resources. A smallholder, by virtue of his small scale, may often be able to farm successfully in a manner defying the commercial conventions of the district, but such success will depend upon his grasp of the principles of soils, crops, and animal husbandry and a wise assessment of his land potential.

3   Getting started

Acquiring the skills

The need for skill in smallholding can hardly be over-emphasized. Farming for a livelihood demands manipulation of the life-processes in the soil, crops, and animals in such a way as to yield continuous sustenance or profit without depleting capital.

Soil husbandry involves much more than mere ploughing and cultivation, although these are hard-won skills in themselves. To till properly, you must know a great deal about every bit of your land – you will find variations between different parts of even one small field – soil texture, structure, acidity, drainage, and history must all be considered; as must the significance of the life within the soil and its relationship to fertility. Not only must you know *how* to plough, but *when*. You must be able to assess your land potential and to farm within it, and know how to build up fertility to reach that potential by the most economic means.

You must understand the habits and characteristics of the crops you plan to grow and how they can best be fitted to the potential of your land. What is more, you must be able to plan your cropping to meet the needs of your animals, your home, and the market, ordering your rotations well ahead to ensure the correct balance. You must know and anticipate your climate and be ever ready to take advantage of any spell of good weather.

You must have a sound knowledge of the life cycles, nutrition, and management of the different species of animals on your farm, and develop this skill to a point where you

can notice, almost instinctively, the slightest signs that a
beast is off-colour. You must then know how to cope with
the trouble and judge whether to act on your own or call the
vet.

You must grasp the complex relationship of all these
subjects to each other; how all the parts fit into the whole. This
is possibly the most difficult of all skills and is a lifetime's
study. The beginner will do well to ponder on it from the start.

The more you can develop these faculties, the fewer will be
your blunders and the better your economy. Even highly
skilled, diligent farmers can misjudge the weather or the
market and make costly mistakes. An unskilled smallholder is
likely to lurch from crisis to crisis.

As well as a deep understanding of your soil, crops, and
animals, as a smallholder you need to be capable of a high
degree of self-reliance in maintaining fences, gates, water
supply, buildings, machinery, and equipment. Unless you can
do so your holding will deteriorate or you will spend a small
fortune paying outsiders to look after it for you. Indeed, you
need to be something of a jack of all trades and passably good
at most of them.

The margins for error on a smallholding are extremely
narrow; only a skilled and diligent husbandman can hope to
succeed. It is therefore prudent for a newcomer to spend
considerable time and effort acquiring the skills before taking
the plunge by buying a farm.

Before you can begin to achieve overall competence you
must master the 'language' to a point where you can think,
and even dream in a new vernacular. The vernacular of any
way of life can only be absorbed 'through the skin' by
immersing oneself in its people; and to this end you would be
foolish not to spend at least a year (preferably two) working on
mixed farms in the district in which you hope to settle, making
friends and winning the respect of the countrymen who will be
your neighbours. It is not easy to find training places on

suitable farms, but it is well worth the search. You should be prepared to work very hard for a pittance. You will feel clumsy and inept at most of the manual skills for at least a year; confidence and speed will only begin to appear the second or third time round. Many skills such as ploughing, using a scythe, working with horses, or developing an eye for livestock may take years to attain; the later in life you come to it, the slower the progress.

Having completed this basic indoctrination, the budding smallholder will know better the joys and frustrations, the rewards and risks, the freedom and the discipline that make up farming. If you then choose to go on with it, you are ready to begin a lifetime of training in depth.

Long-term training in depth may be carried out in many different ways, depending on the character and means of the aspirant. Having learnt the 'language', you can take full advantage of the many facilities available, your ultimate aim being to gain a sound knowledge of the principles of husbandry and become expert in the many practical skills. The day will come when your knowledge of the many-sided principles will come together in an almost intuitive wisdom of the whole, and your skills, transferred from head through heart to hands, will enable you to perform all your physical tasks quickly and well. Only when you have achieved this mastery of theory and practice, will drudgery disappear and the full joy of smallholding be realized.

It is possible to jump in at the deep end, buy a farm, and teach yourself, learning by your mistakes. Such close contact with the facts of life certainly drives the lessons home, but a blunder can be costly, and it is not always easy for a beginner to put two and two together and deduce the root cause of his error. A succession of avoidable mishaps can so sap morale and resources, that it becomes impossible to see the wood for the trees, and your energy is exhausted reacting to emergencies. At such times you need to turn to a wise and experienced adviser who might well have steered you clear of the mistake had he been consulted earlier.

If you have done your practical training on a farm near your own holding, you will have friends and advisers nearby who understand the problems indigenous to the locality. Nevertheless, knowledge and skill can be acquired more quickly through a well-planned training programme, and you would be well advised to consolidate your early practical experience through further training before committing yourself to the full responsibility of your own farm and livestock.

A second year of practical experience on another mixed farm, together with concentrated reading, will confirm and consolidate the earlier lessons; local day-release courses in special aspects of practical farming can fill in some of the gaps. Work alongside competent craftsmen together with such short courses is probably the most sensible path to mastery of the various skills. The more time you have to concentrate on training, the quicker the results will be.

Full-time training at agricultural colleges, universities, or other specialized establishments can provide a much deeper knowledge of the theory of farming, but you should look very carefully at the syllabus to ensure that it covers subjects germane to practical smallholding before enrolling on a course. Some agricultural colleges are beginning to think about organic husbandry and small-scale farming but few have any staff experienced in these aspects and none yet provide full-time courses suitable for a smallholder. Most of their instruction is designed for large, sophisticated farms, and organic husbandry is still usually looked upon as 'muck and mystery'.

Universities tend to train academics and research workers to have very detailed knowledge of relatively narrow aspects of agriculture; their farms are designed for experiment and demonstration rather than to impart practical skills and experience. There are exceptions, however: Reading University now runs an experimental smallholding and Birmingham is doing research into biological husbandry.

The Agricultural Training Board supports a wide range of practical training projects for which farmers and farm workers can form local groups. These courses cover such subjects as sheep-dog training, hedging, ploughing, stone walling, livestock handling, farm mechanics, and, in some areas, working horses. They are extremely practical and instructors are usually skilled local farmers. Having formed a group, the trainees decide among themselves what courses (within certain limits) they need. Any smallholder would do well to join a local ATB group, particularly if he has taken on his own farm with relatively little experience.

A few private training establishments (see pp. 207–10) run long and short courses specially designed for smallholders. So far they have gained little official recognition and therefore do not automatically qualify for government grants to students. For a would-be smallholder who can find the time and money, these probably provide the best possible combination of theoretical and practical training. These centres are the pioneers, the pace-setters for a pattern of training that may, one day, become an integral part of many agricultural colleges. At least two of the centres are registered as charities, and can provide limited scholarships to offset the cost of tuition.

Whatever path you follow to reach your goal of competence, you should make time for reading. No book or course of lectures can substitute for experience, but as you learn the vernacular and gain confidence in your practical ability, you can derive both inspiration and a sense of direction through reading the works of wise men. Most writers have their own pet hobby-horses, and it can be dangerous to swallow the ideas of a single author hook, line, and sinker in the early stages. A guide to suggested reading (see p. 203) can be very helpful in providing a balanced diet.

A small collection of reference books is a most valuable asset on any farm. No one can hope to retain in his head the mass of detailed knowledge needed throughout the farming year. But with a few well-chosen books you can know where to look for the information you need.

I have found, with growing experience, that many of the great authors (Elliott, Stapledon, Scott-Watson, Howard, Balfour, Fraser, and Voisin, to name but a few) yield more and more wisdom at the second and third reading. Like music, the better one knows the score, the greater the delight and reward one gains from reading it.

The innate wisdom of illiterate peasants is legendary. Folklore has been passed down from generation to generation, some of it full of deep and obvious truth, some mere dogma of doubtful veracity. Some genuine truths have been reduced to cant and passed down without explanation. 'It has always been done this way' is poor reason unless one knows why. There generally is a reason for any custom or tradition if only the peasant were lucid enough to explain it. The new smallholder will do well to study local customs and folklore, to question their validity, and try to find out 'why?' before rejecting them out of hand. In the long run, folklore is the accumulated wisdom of generations of practical experience.

Land potential

What precisely do we mean by the term 'smallholding'? A few years ago I was a guest in an Indian village. Asked what I did in Britain I replied 'Like you, I am a smallholder; I farm only ten acres.' 'But that is a large farm; here my family of ten lives on one and a half acres.' By the same token, 500 acres in Saskatchewan might be considered a very small farm indeed. The term is more a measure of the sort of living a holding will provide and the labour involved than a mere acreage. According to the Smallholdings and Allotments Act of 1908, a smallholding was defined as 'an agricultural holding more than one acre and not more than 50 acres in extent, though the area may be extended beyond 50 acres so long as its annual value for the purpose of income tax does not exceed £50'.[1] For the purpose of this book, I should like to think of a smallholding as any parcel of agricultural land of up to (say) 100 acres, organized to be worked by one or two people, without paid labour, and through which they can make part or the whole of their living. At the lower end of the scale, this might be little more than a garden allotment worked in spare time to supplement the income from a full-time job. At the other end it could be a full-time mixed small farm run on commercial lines. The enterprise might vary from vegetables and fruit to livestock rearing, arable crops, or dairying. The prime purpose might be to make money or to achieve a measure of self-sufficiency. The range of possibilities is enormous.

But whatever the size of the smallholding and whatever the intentions of the owner, the first consideration must be the land itself. If the land is poor, no amount of good buildings and equipment will compensate. Poor land means less reward for more work, more risk of disappointment and failure, and it requires greater skill.

It is of the greatest importance for a potential buyer or tenant to assess the potential of a farm before committing himself to it. Only by so doing can he be sure of acquiring a viable holding and be in a position to develop it to achieve its optimum long-term output.

Land everywhere varies in its capacity to produce; output depends on the type of soil and the climate, especially seasonal rainfall, and the limits of production are got by the highest standards of good husbandry . . . In its application to the land *science must be related to inherent land potentialities.* If we fail in this, all scientific discoveries do more harm than good. There is urgent need to determine the limits of production . . . if we are to avoid the danger of trying to extract more from the land than it is capable of yielding. The danger is very real when reclaiming neglected acres. Derelict land may be derelict because some form of husbandry slowly became unremunerative . . . or the dereliction may be the result of exploitation . . . the land being later abandoned. Both areas will be far below optimum today but – and this is the important point – in the one case the land can be brought back to full production easily, whereas in the other case a long and costly process of fertility building will be necessary.[2]

A full and accurate assessment of land potential involves detailed enquiries into soil and climate and experience of the behaviour of the crops and livestock it carries, and this can only be achieved over a period of years. But you can find out a great deal on your initial farm walk, and much more in the weeks and months before you sign a contract or bid at auction. The general appearance of the holding, its grass, crops, and animals, give a valuable first impression. Good, well-managed soil is reflected in the plants and animals it carries. Poor crops and stock, however, may stem from land of low potential or from bad management (or both): it is not so easy to differentiate.

You should be armed with a spade when walking a holding, and examine the topsoil in every field for soil structure and life. The presence of earthworms and deep, strong grass roots are very good signs; shallow, matted roots, a hard 'pan' near the surface, rushes, and surface water indicate a soil in poor structural condition.

Having gained a clear first impression, the next step, before taking the plunge, is to learn much more about the soil itself and the history of the farm. A visit to the nearest Divisional Office of the Agricultural Development and Advisory Service (ADAS) will reveal whether or not the area is yet covered by the Soil Survey of England and Wales. If it is, soil maps can be obtained from the London Map Centre or Sanfords and possibly a detailed 'narrative' of the various soil types, their qualities, and their shortcomings. These maps and narratives can, in some cases, show remarkable soil variations within a very small area and give valuable guidance to their practical management. Time will be well spent talking to the vendor and neighbours from whom you can learn a great deal as to what crops and livestock have been produced over the years, and with what success. From this you can better judge both the potential of the farm and what is involved in bringing it up to that potential.

Buying a smallholding

Having now learned the language and basic skills and knowing how to judge the potential of land, you are well equipped to choose and buy a holding. If your apprenticeship has been in the district where you intend to farm, so much the better, for you will have experience of local conditions and friends to help you look sensibly at whatever offers. However carefully you are prepared, buying a farm can be an exciting and hectic experience, and there is always the danger of making a subjective judgement or acting on impulse; it is here that experienced friends are invaluable.

As stressed before, the potential of the land is of prime importance; if the aim is to farm successfully, all else must take second place. Assuming that you have found land which is potentially viable, your next step is to find out the work, time, and cost involved in realizing that potential. Only then are you ready to consider other factors. **Slope:** what proportion of the farm is too steep to work safely or economically? **Aspect:** will the sun ever be seen in winter? **Drainage:** what needs to be done before a good soil structure can be built up over the whole farm? **Fences and gates:** what will it cost to make the fields stock-proof? **Water:** is there a guaranteed year-round source? Does it supply all the fields? **Access:** can one be sure of getting in and out in all weathers? **Buildings:** are they adequate? What will it cost to make them so? Finally, **the farmhouse:** can it be made comfortable?

Any one of these subsidiary factors may be sufficiently important to render a farm unsuitable, but no combination of them can make a viable farm if the land has insufficient potential.

Farm buildings

The need for farm buildings varies greatly with the type of farming. At one extreme (for instance, when grass-letting) you can get by with little more than the dwelling house: at the other (on an intensively-run livestock farm) special buildings are essential. I make no apology for repeating the fact that a farmer's living comes from the soil and the crops and livestock that grow on it; and not from buildings, machinery, and equipment. It stands to reason, therefore, that outlay on buildings and equipment should be kept to the minimum required for the efficient production of healthy crops and livestock.

Until you have worked out your long-term plans, you have no option but to make the best of existing buildings which may or may not suit your immediate requirements. In the best case you may inherit excellent facilities needing only minor alterations; in the worst there may be no buildings at all or it may be wisest to level the lot and start again. On a typical holding, however, you will use some buildings as they stand, modify others and eventually erect one or more designed to meet your specific needs.

Some buildings, such as milking parlours and farm workshops, are fully used for one purpose throughout the year; others, such as hay barns and shearing sheds are not. Over the years the economics of different enterprises wax and wane: milk gives way to beef, livestock to arable, new husbandry methods are evolved, and new equipment is invented. With each change different buildings and facilities are needed.

When choosing or designing new buildings, simplicity and adaptability are of the greatest importance.

The siting of cowsheds in relation to hay and silage, access to roads and fields, aspects as regards sunlight and prevailing wind, proximity to the farmhouse, and drainage considerations, all deserve careful thought. Well-planned, functional structures, properly sited, can enhance the landscape, particularly if the materials are carefully selected to blend with traditional buildings. An uncoordinated hotch-potch can ruin a beautiful country scene. Once again I stress the wisdom of working backwards, step by step, from the ideal long-term plan into which each new building is slotted in its turn.

The Farmhouse

I have known smallholders who lived in caravans and have seen nomad herdsmen in Asia who camp in the open or carry their tents on their backs. For most of us, however, a comfortable dwelling house is an absolute necessity if we are to do our work happily and efficiently. The style and scope of a farmhouse is a matter of means and taste and is mainly beyond the scope of this book; one point, however, is very important: *the back entrance.* At the best, farming can be rough and dirty work; in bad weather it can be filthy. It is very important to have a well-planned clean-up room beyond which no wellingtons, rainwear or dirty clothing is permitted. Ideally this area should have a tiled floor that can be easily swilled down, plenty of hanging space for rainwear, drying racks for sodden clothing, a wash basin and a lavatory. You should, if possible, have a cold tap and open drain outside the back door to facilitate removal of the worst of the mud from boots and leggings. A clean, smart farmhouse is a great morale-booster, and the provision of good back-door facilities is well worth the cost in the benefit it provides.

Much as you might like to, no smallholder can escape the clerical and book-keeping side of the business. Correspondence and records are best kept in a small office separate from the rest of the farmhouse. In this it is also possible to plan and display future cultivations and other operations.

Having weighed up all these factors, you are ready to make a preliminary capital budget of the costs of fencing, drainage, water supply, access, fertility, buildings, and farmhouse; then of the purchase of livestock, machinery, and equipment. The total of these requirements – some of which will be spread over several years – will give a good indication as to how much you have left to offer for the freehold or rent.

So long as land prices and interest rates remain astronomically high, few will be in a position to buy a holding, however small, with any prospect of their capital paying them a living through genuine farming (as opposed to exploitation or 'mining'). Nevertheless, the real value of land is constant; and capital invested in it will not deteriorate with inflation; in the long term the purchase of a smallholding even at inflated prices, can make sense if the buyer can hold back enough to put it in order, and stock and equip it.

Putting it together

Having bought a potentially viable holding, you must plan a programme for working it up to meet your aspirations. In your mind you will have a picture of your ideal small farm with grass, crops, livestock, vegetables, and fruit meeting most of your needs, without waste, and a marketable surplus to provide a few luxuries. Perhaps you will have a part-time job and a small capital reserve to tide you over the lean years until the farm is pulling its weight.

It is a good idea to start planning from the ideal and work backwards to produce a step-by-step programme over four or five years with clear priorities and objectives for each year. Such a plan will inevitably be modified, but it does provide a framework on which to build.

In the first year you should keep your livestock rate very low; this greatly reduces the chance of disease and ensures adequate winter fodder. It also gives time to improve fencing and drainage, extend water supply and renovate buildings without overtaxing yourself in the early stages. As you gain strength, skill, and confidence you can take on more and more; but if you start off overcommitted, it may be difficult to extricate yourself from an impossible mess.

Looking ahead to the 'ideal' holding some years hence (which, at best, is just a step on an endless ladder), as a new owner you will visualize your farm divided neatly into a number of well-drained, fertile fields, growing crops in a carefully planned rotation to feed both your family and livestock, with sufficient hay and grazing throughout the year. Your livestock will be nicely balanced to make full use of grass and crops without overgrazing the fields and damaging the soil structure. Your buildings, fences, gates, and water supply will be in good order, thus reducing unnecessary work and frustration to a minimum. You will have just enough machinery and equipment to tackle the work without being over-capitalized. The whole picture of crops and livestock will be matched to the potential and location of the holding.

Let us now look back from this idyllic scene and discover the implications. Starting with the imaginary livestock, you soon realize that this population varies with the seasons, so you base your calculations on the breeding cows, ewes, sows, mares, and laying birds, adding their progeny in due season. What are their needs throughout the year – grazing, hay, silage, grain, roots, straw? How much of each can the holding provide? How much land must you commit to arable rotation, and how much must be laid up for hay or silage? What crop rotation will suit the farm? How much feed will you have to buy in? How many secure fields will be needed to cope with such a rotation? What equipment must be acquired? What buildings will you have to alter or build? What drainage and fertility-building must be undertaken to bring the land into full production?

Consideration of these questions will inevitably lead to impossible situations and modification of the original 'ideal'. Gradually, through compromise, a workable plan will emerge, and this can then be broken down into a programme phased over a number of years.

To break the plan into sensible stages, you must have clear priorities. You must build sound foundations before you tackle the upper storeys, and the first of these is the fertility of the land itself. If you plan to keep livestock, the security of fences and gates must go hand in hand with fertility. If drainage is a serious problem, this may have to be tackled before fertility can be built up. It could be ruinously expensive to buy in livestock until you can produce enough fodder; the establishment of a crop rotation and hay or silage-making plans should therefore keep ahead of livestock build-up. Water supply to the fields may be high on the list, although it is generally possible, if laborious, to carry water or move stock to it.

When these top-priority tasks are well in hand, you can start on other highly desirable but less vital works. Farm buildings, the farmhouse, machinery and equipment, orchard and vegetable garden, and the buildup of livestock. It is difficult to see, at first glance, exactly what buildings, machinery, and equipment are needed, or what livestock will best suit the holding. You can struggle by with unsuitable buildings, borrow machinery or hire contractors until your needs become clear. Good neighbours will always help out both with advice and equipment, and at the same time you will be learning about the characteristics of your land and how best to work it.

It may be wise to take over some livestock with the farm, particularly if it is home-bred and acclimatized, but there is a great danger of taking on too much too soon. Possibly the most trouble-free and profitable course in the first year is to grass-let part of the farm, on your own terms, thereby keeping the farm in work whilst leaving you free to get on with your development plans. Grass-letting can be done at auction, by tender, or by private treaty; the last two giving you a better choice of tenant. Lets can be for 364 days or for two separate periods in winter and summer. Another alternative is to buy in store cattle and sheep, selling out in the autumn to clear the land over the winter. Whatever course you adopt, you must avoid overstocking the land and overcommitting yourself in the early stages.

Whilst crop rotations are being established and fertility built up, it will probably be necessary to buy in some feeding-stuffs. This is best done at harvest time when hay and straw can be bought straight 'off the field' and cereals are cheapest.

Fences and hedges

> Little Boy Blue, come blow on your horn,
> There's sheep in the meadow and cows in the corn.

. . . or perhaps worse, your bull has broken into your neighbour's heifers or your lambs are on the road where they could cause an accident. The need for stock-proof fencing is obviously of prime importance.

A sound perimeter (half of which is probably your neighbour's responsibility) should come first, if only to foster good relations; within your boundaries you are then free to work out your own schemes. The second priority will be the division of your holding into a number of main blocks, depending upon your acreage, to give you a measure of control over the grazing of your pastures, to separate groups of stock (e.g. rams from ewes), and to shut off fields for hay or arable crops. Finally, you may wish to subdivide these blocks into smaller areas, either permanently or for short periods for even greater control when strip- or paddock-grazing a ley or for folding stock over roots. The perimeter and main dividing fences must obviously be permanent, and the stronger the better. The rest may be light and movable.

If you take over a farm with poor fences, you will probably feel obliged to stock it before your fencing programme is anywhere near complete. It may well be necessary to 'botch up' repairs using hurdles, bedsteads, corrugated iron sheets, scraps of wire netting, and baler twine. The danger is that you may become so overworked that these stopgaps become a permanent feature of your farm, causing you even more work and worry. You should therefore devise a clear plan with priorities, and introduce your livestock bit by bit as your fencing permits.

By far the strongest and most permanent field divisions are stone walls. Once built they will last forever, the materials for repairs are always on site, and they provide shelter from wind and weather. Few can afford the time and cost of building a new stone wall today, but the principles of dry stone walling, clearly outlined in Rainsford-Hannay's book *Dry Stone Walling*, are straightforward and you can easily acquire enough skill to do effective repairs.

Well-laid hedges of thorn, hazel, holly, or beech (or more generally a mixture of several species) form very effective barriers, and provide shelter for livestock and a habitat for wild life. They require regular trimming to keep the growth thick at the butt (how often one sees huge, bushy hedges with little greenery and gaping holes near the ground) and should be allowed to grow up for relaying every ten to fifteen years. Hedge-laying is a highly skilled craft, and even at moderate charges your hedge will seem expensive. But when you consider the ecological and aesthetic benefits as well as the shelter and livestock control, a good hedge is good value for money. But there are snags: the buds and leaves of hazel and thorn are tasty when grass is scarce, particularly to inquisitive cattle. A single strand of taut barbed wire, $2\frac{1}{2}$ feet high and 4 feet from the hedge is essential to prevent damage and eventual gaps. This is required on your neighbour's side as well.

The most common permanent fence, and the easiest to erect, is the 'stake and wire cattle mesh' (sometimes referred to as 'pig wire') with one or two strands of barbed wire above the mesh. It is my opinion that the heavy grade mesh with the narrower width (6 inches) between verticals is by far the best value in the long run although considerably dearer in the first instance. It is difficult to strain until taut, but provides a much stronger fence. By the same token, your posts, strainers and stakes should always be tanalized (unless you can afford heart of oak). Cattle-mesh fencing is suitable for all livestock although on uneven ground it may be wise to add a strand of barbed wire just above the ground to discourage a rooting pig. The Ministry of Agriculture specifications for grant-aided fencing (available from your local office) give a clear indication of the materials you need.

Strained fences of up to ten strands of high-tensile steel wire can be much cheaper for very long, straight runs. It requires extra-strong, rigid straining posts which are not economic for short lengths. The intermediate posts or stakes are very far apart (hence the relative cheapness of the system) which means that on uneven ground there will be gaps through which lambs may escape.

For temporary fencing you have the choice of hurdles, sheep netting, or three systems of electric fencing. Hurdles of metal, split oak, or wattle are very useful for quick repairs to a hedge or wall or to fold a few sheep on rape or turnips (although they are laborious for the latter). Sheep netting comes in 50-yard lengths and has a 4-inch hexagonal mesh. It is lighter and more flexible than cattle mesh and can be erected fairly quickly on ½-inch-diameter iron stakes which are merely pressed into the ground. It is difficult to get the bottom strained tight enough to prevent a small lamb from burrowing underneath, and the lengths are not easy to roll up neatly after use. Sheep netting has now largely been superseded by various forms of electric fencing.

Electric fencing – single-wire, multi-wire, or plastic mesh netting – is the most common form of temporary fence now in use. It is very quick to erect. For cattle, a single strand is generally enough if the shock is strong enough and the animals properly trained to respect it. For sheep, up to five strands are needed; the smaller the lambs the closer together the wires must be. For pigs, not such agile jumpers, two strands at 6 inches and 1½ feet from the ground should suffice. Electric netting, in which all but the bottom horizontals are electrified, is most effective for sheep. It is no cheaper per yard than permanent cattle-mesh fencing, but is so light and easy to erect that its versatility outweighs the cost.

All electric fencing requires regular attention to ensure the system is 'alive' and not earthed out by wet grass touching the wires. Netting is particularly prone to such earthing, and the bottom horizontal (which is not electrified) is frequently bitten through by rabbits or badgers, making gaps for lambs to squeeze through. I have known a sow that enjoyed the sensation of a pulsating electric shock and a cow that learned to uproot a mesh fence with her horns. Dry batteries for fencer units are outrageously expensive, so I have resorted to an old tractor battery, trickle-charged by a windmill made from a bicycle wheel with its dynamo hub: it cost nothing and worked without trouble for a year at a stretch.

However carefully you maintain and repair it, electric fencing deteriorates, and, apart from the electric pulse units, must be replaced every six or seven years.

Suiting enterprise to holding

Many factors must be taken into account when choosing the most suitable enterprise for a smallholding; some fixed, others variable and within the influence of the farmer. Soil and climate, which should be taken together, determine the land potential (see Chapter 2) and largely dictate the main types of farming – arable, stock-rearing, fattening, dairying, etc. – that are most likely to be viable. Altitude, aspect, and slope may further limit the scope. If the farm has hill-grazing rights, these may assume an overriding importance in relation to other factors. The size of the holding and its proximity to a town and markets have obvious implications and are beyond the control of the occupier.

Possibly the most variable factor is you, the smallholder. What do you aim to achieve? A cash income? A measure of self-sufficiency? A full-time way of life or a spare-time balance to another occupation? Do you have a delight in animals and skill in their management? Are you a jack of all trades or do you prefer to specialize? Do you and your family have any particular skills to exploit?

Within the limitations imposed by the 'fixed' factors, you have a wide choice before you. The larger the holding and the stronger the soil, the wider your choice.

Let us first consider the 'commercial' smallholder for whom money-making is the overriding aim. If we discount the unscrupulous operator who is prepared to 'mine' his land for a quick profit, we are considering a responsible husbandman who aims to make a cash living whilst conserving and improving his capital, the soil. As a commercial smallholder you will relate all your enterprises to the market, which assumes a dominant influence over your choice. You must either study and cater for existing markets or create new markets for your own special products. Existing markets are very competitive and, especially in poultry and pigs, are dominated by agribusinessmen. Unless you can sell top-quality produce at a premium or, through hard work and efficiency and the 'personal touch' can get high production at low cost, you will be hard put to compete against larger farmers whose 'per-acre' capital investment is generally so much lower.

First amongst small commercial enterprises must be the dairy herd, where the personal touch of a skilled stockman can raise production, reduce costs, and maintain good health and longevity in cattle far beyond the scope of most large herds. But cows must be milked twice a day, every day of the year, so if you are a dairying smallholder you must find your recreation in the variations within the farm. A great advantage of milk is the regular cash income. If you sell farm butter and cheese instead of whole milk you can do well if you have the right market. This is a labour- and time-consuming enterprise if done on a commercial scale.

Pig-keeping, too, responds to individual attention, although the market tends to fluctuate rather widely. Pig enterprises can vary from outdoor breeding and weaner-rearing to intensive indoor pork or bacon production. The price of bought-in feeding-stuffs and the capital required for intensive pig-keeping can be limiting if not prohibitive factors for a commercial smallholder. If a regular source of swill is available nearby, pig-keeping can be profitable, but health regulations impose severe restrictions on swill feeding.

The breeding, rearing, or fattening of beef cattle can be viable in the right conditions, although it is hard to make a profit on a very small scale. Here there is a choice of buying in cattle and selling later – a 'flying' herd – which has the advantage of relieving the land of excess livestock in winter.

With hill-grazing rights, there is a place for sheep on a smallholding. Without it, in my opinion, the profit does not compare with that of dairy cattle. But, as a means of maintaining good swards, it is a good policy to mix sheep with cattle.

A well-run herd of goats can be profitable near a town where a specialized market can be established.

There is a growing premium market for free-range eggs and table birds which can provide a profitable enterprise for an expert small-scale poultry man. Unless you can find such a market, there is little prospect for small-scale poultry-keeping in competition with the huge multinational businesses. As with pigs, the high cost of bought-in food is a major deterrent to success. The scratching of poultry in fields and the value of their droppings add an unmeasured profit.

Vegetable and fruit-growing for a nearby market can be profitable on the right land, but this is very labour-intensive, and does not combine well with stock farming, which also requires the undivided attention of the smallholder. Nevertheless, sustained horticulture needs an abundant supply of organic waste to maintain soil structure and fertility. The growing demand for genuine, organically-produced vegetables and fruit offers a profit to the skilled gardener who is near enough to his market.

Bee-keeping can be a profit-making project, although in areas of high rainfall, results can be disappointing. In conjunction with garden and orchard there is an added bonus in the better fertilization of fruit trees and garden produce.

Bearing in mind the long-term structure and fertility of the land, most smallholders will generally combine two or more of the above enterprises, and try to grow most of their own food. The temptation is always to overstock, and this means that the land is asked to produce more than its true potential. In the long run fertility and crops decline and less stock can be carried. It is a fallacy to think that manuring (either organically or artificially) can overcome the effects of overstocking except in the short term. On the other hand livestock is the source of fertility. Pigs and poultry, sensibly managed, can be a rapid means of building up fertility on a run-down holding.

Should your prime aim be to provide a measure of self-sufficiency for your family, a different range of options opens to you. Of course you are still bound to make ends meet, but you no longer have to make all your decisions in terms of money profit as the first requirement.

For a start, you will consider the needs of your family: bread, meat, dairy products, eggs, vegetables, and fruit – most, if not all, of which can be produced on a smallholding. Fuel and clothing will probably be more difficult, and some needs can only be met by buying in from outside. There must, of course, be money-making enterprises as well to buy those necessities that cannot be grown on the farm, but these will be looked upon as secondary calls on the agricultural operations in which farm surpluses are sold to supplement the cash income from non-farming sources.

Depending upon the size of your holding, you will keep one or two house cows or goats, turning their milk into butter, cheese, and yoghurt, and possibly a few beef cattle as money-spinners. You may fatten a couple of pigs to kill and keep a small flock of laying hens or ducks on household scraps. You will have a small vegetable garden and orchard and probably keep bees.

If you have a spinner in your family, you will keep some sheep; possibly with fleece rather than carcase quality determining the breed.

The larger the holding, the greater the surplus after family needs are met, and the more you can think of commercial enterprises as money-spinners. In the end, it is your own 'delight', capability, and strength that will limit what you undertake. At some point the size of the enterprise may dictate that you think commercially rather than in terms of self-reliance. If you can keep your 'surplus' projects simple and labour-saving – grass-letting or flying flocks of grazing animals for example – you may still find time to concentrate on feeding your family as your first objective on quite a large holding.

Figure 2 shows the layout of my own ten-acre farm. The land is old red sandstone with a very high content of fine silts. The house is at 620 feet above sea level and the farm is on a gentle east-facing slope.

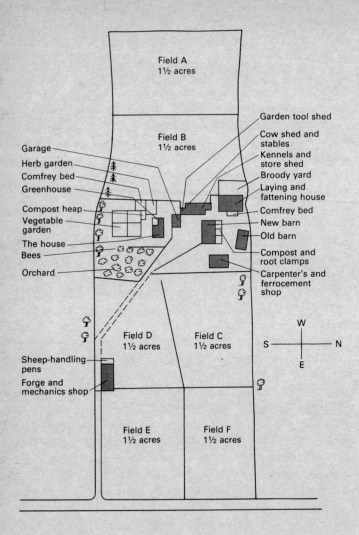

Field A
1½ acres

Field B
1½ acres

Garden tool shed
Cow shed and stables
Garage
Kennels and store shed
Herb garden
Broody yard
Comfrey bed
Laying and fattening house
Greenhouse
Comfrey bed
Compost heap
New barn
Vegetable garden
Old barn
The house
Compost and root clamps
Bees
Carpenter's and ferrocement shop
Orchard

Field D
1½ acres
Field C
1½ acres

W
S — N
E

Sheep-handling pens
Forge and mechanics shop

Field E
1½ acres
Field F
1½ acres

Fig. 2. Sketch map of the author's farm

I have found a seasonal calendar, as shown in Table 1, a useful reminder of the sequence of work through the year.

The initial budget

So far we have considered our capital outlay on land purchase, land improvements, livestock, machinery, equipment, and buildings, and decided how this will be spread over the 'build-up' years. You must now determine what financial help you need (including Government subsidies for such things as new buildings, fencing, hedging, hill cattle, and sheep), what it will cost, and how it is to be repaid. A detailed planning budget for both capital and operating accounts must now be prepared not only to give you the information you require for your own plans, but also as the main evidence to persuade the Bank Manager to finance your endeavours.

In Table 2 I have shown a typical capital budget over four years. The figures are based on my own in the 1970s and now bear no relation to the present day, but they serve as a broad outline.

Having worked out the capital expenditure and borrowing requirement, you can determine what surplus income will be needed to repay the loans, plus interest, over a given number of years. It is not difficult to see that a smallholding cannot support a large mortgage as well as a family!

Now we come to the current account – the year-by-year income and expenditure on which the holding will succeed or fail. At this point cast your mind ahead once again to your ideal farm (in, let us say, Year 4, to be optimistic) with crops and livestock nicely balanced to suit the land potential. Now you can estimate the value of produce, for sale or home consumption, for a full year and balance this against likely expenditure.

In Table 3 I show a one-year budget, based roughly on my own farm in Year 4. The individual figures are less important

Table 1. Farming year calendar as used on the author's farm

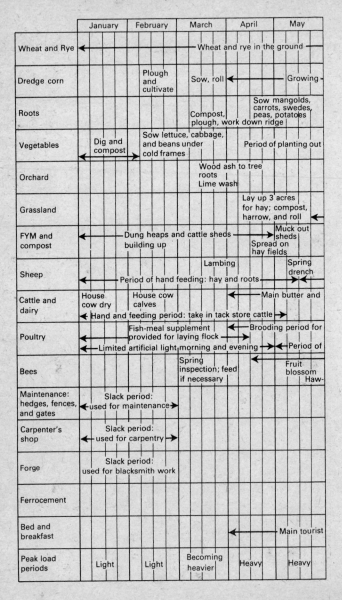

	January	February	March	April	May
Wheat and Rye	←		Wheat and rye in the ground		→
Dredge corn		Plough and cultivate	Sow, roll ←		→ Growing
Roots			Compost, plough, work down ridge	Sow mangolds, carrots, swedes, peas, potatoes	
Vegetables	Dig and compost ←	→	Sow lettuce, cabbage, and beans under cold frames	Period of planting out	
Orchard			Wood ash to tree roots / Lime wash		
Grassland				Lay up 3 acres for hay; compost, harrow, and roll ←	
FYM and compost	←	Dung heaps and cattle sheds building up		→ Spread on hay fields	Muck out sheds
Sheep	←	Period of hand feeding: hay and roots	Lambing	→	Spring drench ← →
Cattle and dairy	House cow dry	House cow calves		← Main butter and	
	← Hand and feeding period: take in tack store cattle →				
Poultry	←	Fish-meal supplement provided for laying flock →		← Brooding period for	
	← Limited artificial light, morning and evening →				← Period of
Bees			Spring inspection; feed if necessary	←	Fruit blossom / Haw-
Maintenance: hedges, fences, and gates	Slack period: ← used for maintenance →				
Carpenter's shop	Slack period: ← used for carpentry →				
Forge	Slack period: used for blacksmith work				
Ferrocement					
Bed and breakfast				←	Main tourist
Peak load periods	Light	Light	Becoming heavier	Heavy	Heavy

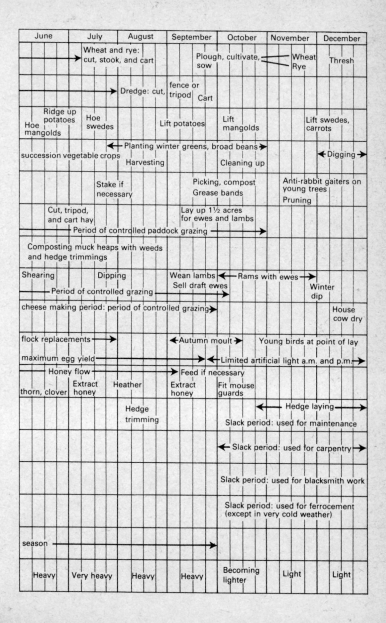

Table 2. Capital budget

Item	Year 1	Year 2	Year 3	Year 4	Total
Expenditure (£)					
Purchase of farm	25,000	—	—	—	25,000
Fences and gates	800	400	—	—	1,200
Livestock – cattle	600	300	—	—	900
sheep	300	150	—	—	450
pigs	—	—	200	—	200
poultry	50	50	—	—	100
horses	—	—	600	—	600
bees	—	—	150	—	150
Dwelling house	3,500	—	—	1,200	4,700
Buildings and fixed equipment	1,500	—	—	2,500	4,000
Machinery and tractor	500	1,500	500	500	3,000
Tools and workshop	150	150	100	—	400
Drainage	—	—	1,500	—	1,500
Water supply	—	450	—	—	450
Access road	—	—	800	—	800
Totals:	32,400	3,000	3,850	4,200	43,450
Accumulative:	32,400	35,400	39,250	43,450	
Income (£)					
Sale of timber from forestry plot	—	4,000	—	—	4,000
Net capital required (£)	32,400	31,400	35,250	39,450	

than the overall picture, which illustrates the limitations of production on a small farm and the unavoidable expenses. It also emphasizes the need for 'money-spinning' enterprises if the smallholder is to rise above a very bare subsistence living *and* pay off a mortgage at the same time.

Once you have established a clear picture of your ideal (and realized its limitations) you work back step by step to the beginning and prepare a priority list for your development plan. It may be necessary to adjust your ideal to reconcile it with reality, and prepare two or three budgets before you go firm on a viable plan.

I have found it very useful to prepare a graphical cropping

Table 3. Typical one-year budget

Income (£)
(including produce consumed on farm)

Item		£	£
Cattle (1 house cow, 1 nurse cow)			
	Milk	30	
	Butter	150	
	Cheese	300	
	Calves	300	
	Skins	10	790
Sheep (10 breeding ewes)			
	Draft ewes (3)	60	
	Lambs (10)	180	
	Wool	25	
	Skins	25	290
Horses (1 brood mare, 1 riding pony)			
	Sale of hunter gelding	500	500
Poultry (45 laying birds)			
	Eggs (750 dozen)	300	
	Table birds	50	
	Breeding stock	25	375
Vegetables			50
Fruit			50
Honey			100
Bed and breakfast			500
Work for neighbours			100
Stock on agistment			120
Government subsidies			450
Total income			**£3,325**

Expenditure (£)

Item		£	£
Bought-in fodder			
	Hay	250	
	Straw	160	
	Oats	150	
	Mixed corn	160	
	Dog meat	100	820
Seeds			60
Fertilizers			100
Repairs and renewals – Fences and gates			
	Buildings		
	Implements		500
Tractor – Fuel			
	Tax		
	Insurance		
	Repairs		
	Depreciation		250
Motor (share of total 1,000)			
	Fuel		
	Tax		
	Insurance		
	Repairs and service		
	Depreciation		300
Light and heat (share of total 300)			100
Post and telephone (share of total 150)			100
Veterinary			50
Contractors – baling etc.			100
Haulage			20
Insurances (not car or tractor)			100
Financial costs – accountants' fees and interest on overdraft			100
Total expenditure			**£2,600**
Net profit			**725**
			£3,325

Figures are rounded to the nearest £5

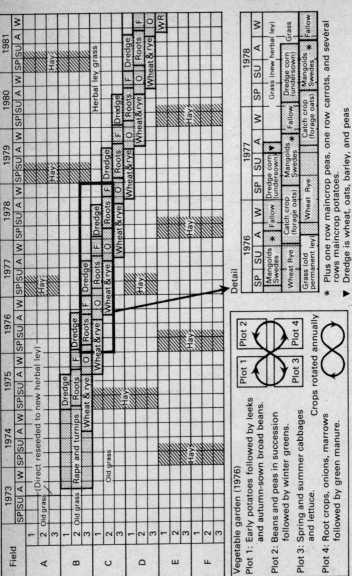

Fig. 3. Crop and vegetable rotations: arable master plan as operated on the author's farm

The land is divided into six fields each of roughly 1½ acres. Each field is temporarily subdivided into three ½-acre sections for the crop rotation. Root crops are ¼ acre of swedes and ¼ acre of mangolds each year, plus one row of maincrop peas, one row of carrots, and several rows of maincrop potatoes for domestic use.

plan (see Figure 3) looking ahead a number of years. On this you can balance extractive crops with fertility-building leys, grazing with cutting, and thereby keep the individual fields in balance with the overall potential of the land. At the same time you can plan your grazing and crops to meet the needs of the livestock.

But while recommending such long-term plans, I must stress the need for flexibility. No two years are alike, and what succeeds this year may fail next: further, your early plans will be made without complete knowledge of all the factors, and modifications are bound to be necessary. Nevertheless, a basically sound plan provides a framework on which you can build and progress in the light of experience and changing conditions.

The aim should be flexibility within a structure; variations on a theme.

4 The living soil

The foundations of health

Whatever the commercial motivations of those engaged in the 'farming industry', the ultimate purpose of agriculture must be to provide mankind with a sustained and ample supply of health-giving food. As the world population grows, our limited resources, particularly of land, will have to be used efficiently and frugally to achieve this paramount aim. As I have written elsewhere:

The late Sir Robert McCarrison determined very clearly the qualities of food required for sound nutrition; unadulterated food, unrefined food, fresh food, raw food. Yet he went further than this ... food must be produced from a healthy, living soil. It is not merely a matter of the known constituents of food; fibre, protein, carbohydrates, fats, vitamins, minerals, and trace elements in the correct proportions. Animals can be sub-clinically sick and still be eaten by man. Plants can be sick, yet fed to both animals and man. The soil, itself a vital symbiosis of plant, animal and mineral, can be sick, yet grow plants to feed animals and man.

As long as man eats unhealthy food (the product of unhealthy soil), even in the right proportions, the medical profession will perforce have to continue to treat the inevitable results. As long as animals eat unhealthy plants, the vets will have to do likewise. As long as plants are grown on sick land, the chemists, plant pathologists and entymologists will be forced to continue with their pesticides and herbicides, with escalating side effects ... not least of which is the further sickening of the soil.

So, in the end (or should one say 'at the beginning'?) it comes down to the health of the living soil.[3]

The origins of soils

Soil consists of a mixture of mineral (rock) particles, organic debris, and living organisms permeated by air and water. It is a dynamic system of continuous progressive changes due to external environmental conditions and to the processes of synthesis and decay brought about by the living organisms within the soil itself. The mineral fraction of soil consists of a wide range of particles from boulders through gravel, sands, and silts to clays.

In Britain, the gravels, sands, and silts have been produced by the breakup of rock by physical forces, chief of which was the grinding action of the Ice Age. The particles produced in this glacial mill vary in size from rocks to silt. The sand and silt particles, which are roughly spherical in shape range from 2,000 microns ($\frac{1}{16}$ in.) to 50 microns in diameter (sands) and from 50 to 2 microns (silts); the finer silt particles being indistinguishable individually to the naked eye.

Clays, the product of *chemical* weathering, have very much smaller particles (one particle of fine silt would make 1,000 particles of clay) which are flat or plate-like in shape and lie against each other like sheets of glass, sticking tightly together as water is drawn between them.

The organic matter in soil consists of plant and animal debris or waste in varying stages of decay. The proportion may vary from 100 per cent (peats) to practically none (sand dunes).

Soil texture

Soil texture has been defined as 'a store of building materials' – stone, gravel, sand, silt, clay, organic matter, etc. – and is a measure of the potential of the land concerned.[4] The pores between silt particles are so small and retentive of water that tightly packed silt cannot be emptied by drainage. Sand, however, drains easily as the pores are so much larger. Silt can lift water several metres by capillary action, sand can raise it about 30 cm., but stones and gravel cannot lift it at all.

However tightly packed together, pure sands will allow the free passage of water and will not form 'pans' or 'poach' even when mistreated. Nevertheless, they cannot, of themselves, form the spongy crumb structure required to retain moisture, and therefore tend to be either water-logged or completely dry.

Because of their tendency to pack and hold water, silts are frequently mistaken for clays. In fact they are very different, due to the size and shape of their particles as discussed above. Both silts and clays can be heavy and difficult to work, but require completely different treatment.

When water is drawn between the plate-like clay particles, they stick tightly together and swell. As clay dries out it shrinks, causing cracks and a fine granular tilth. By contrast, wet silt tends to become a 'porridge' which dries out into large clods with fewer, large cracks. When these dried clods are broken down by tillage, they tend to disintegrate into 'flour', having little of the cohesive strength of clay.

Clay particles carry a negative electric charge enabling them to hold many important positively charged plant nutrients such as calcium, magnesium, potassium, and nitrogen (in the form of the ammonium ion). Indeed negatively charged clay particles act as bridges between these minerals to form flocculated clay 'packets' which are stable against mineral leaching.

Organic matter can completely replace the mineral fraction of soils in agricultural production as is the case in the fen peats or in organic greenhouse composts. Once it is properly drained and aerobic, peat can retain mineral nutrients and trace elements in readily available form even better than clay. It is a slow-release source of nutrients and provides large pores for drainage and root penetration and small pores for water retention. On its own, however, it tends to need 'ballast' in the form of sand or loam, which provides mechanical stability to prevent plants from being blown out of the ground.

Apart from the peats which were formed by the anaerobic decomposition of plant life in waterlogged conditions, organic matter in soils consists of a wide variety of living animals and plants, including earthworms, insects, bacteria, plant roots, and fungi, as well as decaying animals and vegetable fibres and wastes. As these wastes are broken down by physical, chemical, and bacterial action, they gradually form 'humus', a highly important binding factor in the formation of stable soil crumbs or tilth and in the transformation of minerals into soluble plant nutrients.

Soil structure and fertility

We have likened soil texture to the store of building materials available and the measure of land potential.[4] Soil structure is what nature and man make of these materials; living room, dining room, water supply, ventilation, and drainage. To build the most stable and convenient soil structure possible from the available resources should be the most important task of every farmer.

The ideal soil structure provides a firm and stable 'bed' in which plants can stand and grow. This bed will contain abundant stores of water and plant nutrients readily available at all times. It will be free-draining so that it is never waterlogged and will be well aerated. It will be of good depth to encourage deep rooting and exploitation of its resources. This perfect plant habitat will further be strong enough to withstand the physical punishment inflicted by the weather and by the animals and machines that harvest the crops it bears. Finally it will contain the facilities to repair and rebuild itself should any combination of climate, beast, and machine conspire to break the structure down.

The method used to build the most stable and convenient

structure possible will vary with the soil texture and climate. Properly understood and managed, strong clays will structure themselves, and can nourish plants with relatively little organic matter present. If pure sands can be given sufficient organic matter to retain moisture and provide plant nutrients, they need no further structure as the pores between sand particles are large enough to allow drainage and sand cannot be compacted into a 'pan'. If peats can be drained, sweetened (brought to a neutral pH) and kept mechanically strong enough to support plants, they are complete in themselves, although when ploughed they can oxidize rapidly into non-existence. The almost pure loessic silts are incredibly fertile, but are by far the most unstable and difficult soils to work. On their own they lack the ability to form large pores for drainage and aeration, and are most vulnerable to wind erosion, panning, and poaching.

Owing to the action of the glaciers, the majority of Britain's soils are mixtures of sands, silts, and clays. The relative proportion of silt to clay is remarkably uniform at about six to four; it is the sand fraction that varies most. Few soils contain the 40 per cent clay to qualify as true clays; few are pure silts or pure sands. The common mixtures of sand, silt, and clay are called loams (silty loams, sandy loams, clay loams, silty clay loams, and so on) and in the absence of organic matter, reflect structural qualities in proportion to their ingredients. The introduction of organic matter can change everything by the formation of a 'crumb' structure in which large pores allow aeration, root penetration, and drainage, and small pores (within the crumbs) retain water and nutrients (see Figure 4). It is the building and maintenance of this crumb structure or tilth that forms the basis of soil husbandry.

The adequate return of organic wastes to the soil through its own decaying vegetation, grazing animals, rotted farmyard manures, and the avoidance of continuous extractive cropping is the primary measure required to maintain fertility. But this

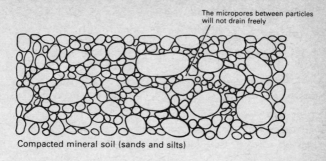

Compacted mineral soil (sands and silts)

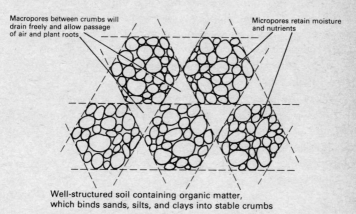

Well-structured soil containing organic matter,
which binds sands, silts, and clays into stable crumbs

Fig. 4. Soil structure

would be to little avail if the ground were waterlogged or so
acid as to prevent the normal action of soil life. To realize the
full potential of his land a farmer must enlist the support of
many allies by providing conditions in which they will thrive.
These include various types of mycelium (fungi), bacteria,
and, most important, burrowing earthworms. This soil life, in
proper balance, constitutes the digestive metabolism of plants
through which the raw mineral and organic nutrients are
combined into soluble, 'available' forms. In many cases the
threads of mycelium invade the plant roots and act in

symbiosis as channels for nutrients between soil and plant. For growing crop plants the following conditions must be present:

1. The first essential of healthy soil life is air, and this means good subsoil drainage and adequate inter-crumb pores. Until drains are put right, other factors, however favourable, are nullified.

Equally important is moisture; neither too little nor too much. This is guaranteed by a good crumb structure with fine pores holding water like sponges, whilst allowing air and surplus water to move freely between the crumbs.

2. The second requirement is a more or less neutral soil; neither too acid nor too alkaline. Most of the useful soil flora and fauna (particularly earthworms) do not thrive in very acid soils (below pH5). A neutral soil of pH6 is ideal, and the addition of ground limestone, calcified seaweed, slaked lime, and/or basic slag may be needed to achieve this.

3. The third need of soil life is food, which can be provided by the return of organic waste in the forms of rotting vegetation from the plant cover, farmyard manure, compost, or from green crops specially grown and ploughed in. The grass-ley break in an arable rotation is a typical attempt to provide food for life in the soil.

4. Finally, having achieved these conditions, it is most important the resulting structure should not be destroyed mechanically by the wheels of heavy machinery or the feet of livestock in wet weather.

The provision of these conditions will encourage the buildup of soil life, particularly burrowing earthworms, and it is this active soil population which serves as the 'repair gang' to build an increasingly stable structure and to repair damage as it occurs. At the same time, the earthworms burrow deep into the subsoil bringing minerals to the surface, churning and mixing mineral and organic matter in their gizzards, depositing the excavated material on or near the surface in very stable crumbs exactly the right size for optimum exploitation by plant root-hairs. Their burrows act as ventilator shafts, drain-pipes, and root channels.

For the most part, soil life thrives in the top six inches of the earth where moisture, air, nutrition, and temperature are most amenable. Below this is the subsoil which is relatively inert. Deep ploughing puts subsoil on top and buries the living topsoil below the optimum depth. Ploughing serves two purposes: to bury turf and weeds, eliminating competition with the desired arable crop, and to stir and mix mineral and organic fractions of the topsoil. These two aims conflict, and it can be difficult to strike the ideal balance; but to build and maintain good soil structure and fertility one should avoid ploughing where surface cultivation will meet the need, and abandon deep ploughing (over six inches) altogether.

Soils may be classified under two main headings: brown earth and podzol (see Figure 5).

A brown earth soil is aerobic, of more or less neutral pH, with a good mixture of mineral and organic matter, bearing deep-rooting plants and supporting a normal population of soil life, particularly earthworms. A podzol soil is liable to waterlogging, is acid, organic

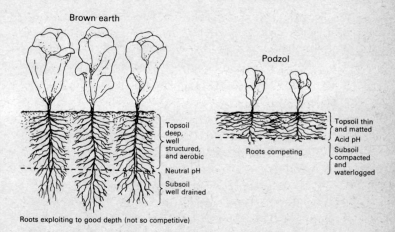

Brown earth

Podzol

Topsoil deep, well structured, and aerobic

Neutral pH

Subsoil well drained

Topsoil thin and matted

Acid pH

Subsoil compacted and waterlogged

Roots competing

Roots exploiting to good depth (not so competitive)

Fig. 5. Typical plant root development in brown earths and podzols.

wastes remain unrotted near the surface; the topsoil is thin; root penetration is shallow and matted. Soil life is scarce and burrowing earthworms absent. Normal grasses, clovers, and arable crops require brown earth soils; heathers, rushes, and coniferous trees have adapted themselves to thrive in podzol conditions. To change a soil from podzol to brown earth may be both laborious and expensive, and will involve the provision of the four conditions mentioned earlier. When reclaiming such land, it may be necessary to provide the first pioneer crops with a direct injection of nutrients through chemical fertilizers but it should always be remembered that the aim is to build up the life in the soil so that it can provide its own fertility within the land potential. Concentrated sources of soluble nitrogen applied on the surface as superphosphate or sulphate of ammonia encourage shallow rooting and discourage earthworms. If the burrowing earthworm population can be sufficiently built up, soil fertility (within the limits of its potential) will take care of itself. Without earthworms and other soil life, no amount of cultivations and chemical fertilizers will build *and sustain* true structure and fertility.[5]

Preparation and use of organic wastes

The natural cycle by which organic wastes are returned to the soil and broken down into humus presupposes a balance between soil, plants, and animals. There is no 'extractive' cropping and no import of fertility from elsewhere. When man comes into this cycle, he brings with him the possibility of radical changes which can completely upset the natural balance. He can fell forests, drain swamps, introduce and control new crops and livestock, import fodder and fertilizers, and export crops, animals, and milk. Not only can he build up and improve the fertility of his land, but also exploit that fertility up to and beyond its latent potential.

In nature the cycle is practically closed, with vegetable and animal wastes returning directly to the soil where they have lived. All this waste is attacked by a multitude of living organisms which progressively break it down and 'digest' it into humus. The bodies of dead animals are devoured by other creatures large and small whose excreta and remains are, in their turn, returned to the soil. Animal dung is spread by birds, devoured by insects, bacteria, and earthworms and mixed intimately into the topsoil. The leaves of trees, grasses, and plant roots return direct to the soil or through the guts of grazing or browsing animals. Thereafter the process of breakdown takes place entirely on or just below the surface of the earth. The speed of breakdown depends upon the composition of the wastes, the availability of air and moisture, temperature, and the presence of the right soil fauna and flora. A well-balanced mixture of vegetable matter (carbon) and animal waste (nitrogen), deposited in warm, damp conditions on a soil full of earthworms, bacteria, and fungi, will rot down very quickly.

This matter of cellulose breakdown is of vital importance to the whole question of the nutrition of the living cells, for it has been shown by experiment that raw cellulose can be injurious to plants as it is indigestible to animals, and in both the soil and the digestive tract of herbivora, cellulose decomposition is largely performed by fungi.

Such soil treatments, therefore, as ploughing in straw or green crops can only prove successful if conditions for the subsequent breakdown of the cellulose are present. It is useless to apply organic matter in this form while at the same time inhibiting, by the use of certain artificial fertilizers, the fungi capable of converting that organic matter into humus.[6]

Author's note: The injurious effect on plants refers to the fact that a surfeit of undigested cellulose upsets the C/N ratio and delays the breakdown of organic matter, thus causing a slowing down in plant growth.

Stemmy plant fibres (cellulose and lignins) in cold, drought, or waterlogged conditions will take much longer, and will *retard* plant growth until they have broken down. By intensifying and controlling his livestock and introducing arable crops, the

farmer must plan carefully to ensure that the natural cycle of
return continues to operate evenly over the whole of his land.
Cereal-growing or hay- and silage-making in the same field
year after year will obviously rob that field of humus and
structure, especially if heavy-yielding crops are grown with
chemical fertilizers. By the same token, permanent grassland
that is overgrazed by growing youngstock or milch cows and
never cut will become deficient in phosphorus and calcium
and may lose structure from poaching. In such cases the
addition of organic matter, slag, or lime will be essential if the
farming system is to be maintained without a drastic drop in
fertility and production – or the breakdown of soil structure
which ultimately leads to the same thing.

Lime (to keep the pH near 6 and replace the calcium sold
off in milk) and basic slag (to replace the phosphorus and
other elements exported in animal bones) can be applied by
contractor relatively cheaply to all but the steepest fields.
These minerals are relatively insoluble and act not so much as
direct plant foods, but put the soil in a state that favours the
buildup and activity of soil life. Nevertheless basic slag is an
artificially produced fertilizer, and its acceptance by organic
farmers who reject 'artificials', poses a dilemma. It is enough to
say here that slag should be used sparingly as a soil conditioner
from time to time, and not as a routine crop food. Nitrogenous
chemical fertilizers *are* direct plant foods which bypass the
normal metabolism of soil fauna and flora, and in some cases
inhibit their reproduction and useful activity. They stimulate
vigorous competitive root growth near the surface and can
force plants to grow too quickly to take up their full needs of
other nutrients. In strong soils these side effects may take some
years to show up; on poor land they can exhaust the humus
very quickly with disastrous results. That they can increase
production in the short and medium run is beyond doubt, but
unless employed in the full knowledge of land potential and
soil life they can soon lead to diminishing returns.

There are many sources from which the farmer can obtain organic wastes for his soil, some from without and others from within his farm. If he farms near a town he may be able to buy, still very cheaply, municipal compost and sewage sludge, ensuring beforehand that they do not contain poisonous mineral residues such as lead. Fish-meal (with or without added synthetic NPK), dried seaweed, liquid seaweed extract (Maxicrop), guano, bone-meal, hoof and horn, dried blood, and shoddy may also be purchased, but are now costly. Bought-in feeding concentrates for livestock are an indirect fertility import. On occasions one can buy cow, horse, poultry, or pig manure from neighbouring farms or stables, but few farmers can now afford to part with such valuable by-products.

From within the farm, organic matter can be returned direct by grazing animals; as liquid slurry from dairy cattle sheds; as raw or rotted farmyard manure (FYM) consisting of dung, urine, and bedding; as rough or fine compost; as a specially grown green manure crop ploughed in; or as a grass ley.

It has long been known that raw, unrotted FYM does not produce a quick response in crops because the bacteria have to work so hard to break it down, and that fresh pig or poultry dung is too 'hot' for young plants. It has thus been customary for dung and bedding to be rotted for up to a year in a midden (or field heap) before spreading on the land. A farmyard midden may become waterlogged, its decomposition anaerobic, and substances may be leached out by rain. Dung and bedding are added, as they come, with no particular thought to the balance between carbon- and nitrogen-rich constituents. The heap may or may not heat up due to aerobic bacterial action, and the end product may be more or less valuable accordingly. Slurry may be mixed with water and spread by tanker or irrigation pipeline. Considerable storage space in tanks or lagoon is required to enable spreading to be done only when ground and weather conditions are right.

Compost-making: the general principles

Properly made compost is greatly superior to FYM or slurry. In the first place compost is made from a carefully balanced mixture of animal wastes (the nitrogen fraction from which the soil bacteria are built up) and vegetable matter (the carbon fraction which provides the energy with which the bacteria heat and break down the heap). The end product is friable humus which is ready to feed the soil flora and fauna and, through them, the plants. Even rotted FYM must be further broken down before it is fully used. Raw muck and slurry can burn young plants and even slow down plant growth whilst it is being broken down in the soil. Secondly, well-made compost preserves all its constituents in a stable form, whereas FYM is liable to lose a great deal of its value through leaching and evaporation. Finally, there is strong evidence to show that humus in the soil can confer a remarkable degree of resistance to plant and animal pests and diseases.

Compost-making on a farm scale is yet uncommon in Britain (it has been practised for over 4,000 years in China) as it involves a very careful balance and mixing of ingredients, control of ventilation and moisture, and the extra work of at least one turning. The prospect of setting yourself up for this job is far more daunting than the actual work, and once you have overcome the inertia of thinking it out and making a start, the results more than justify the effort. Not only is the end product more valuable than slurry or FYM, but it can be spread much thinner (7 or 8 tons to the acre as opposed to 15 tons of FYM).

The principles and practice of compost-making are described in great detail by A. Howard, M. E. Bruce, and others. The 'Indore' method as developed by Howard is probably the most practical for farm purposes. In essence the process consists of mixing vegetable and animal wastes, adding a base to neutralize acidity, and managing the mass so that the aerobic micro-organisms which break it down can work to the best advantage. This involves control of moisture, ventilation, and retention of heat.

Vegetable wastes include straw, damaged hay, hedge trimmings, nettles, lawn mowings, weeds, potato haulm, bracken, leaves, sawdust, wood shavings, paper, and kitchen waste. Particular green crops such as comfrey may be grown specially for compost heaps.

Animal residues include the urine and dung of domestic stock – poultry droppings, pig, cattle, sheep, and horse dung – dried blood, slaughterhouse refuse, powdered hoof and horn, fish manure, sewage sludge, and shoddy.

A base is required to neutralize the acidity that naturally builds up during fermentation. Ground limestone, calcified seaweed, or wood ash may be added to keep the pH near 7 for the optimum action for micro-organisms. Slaked lime is less effective and quicklime is altogether too fierce.

The process must be kept aerobic and moist in the early stages. If too much water is added, the process becomes anaerobic too soon; if too little water is present, bacterial activity slows down and ceases. It is therefore important that the heap is built of layers of vegetable and animal wastes, evenly distributed and not compressed or trodden to a point where air is excluded. In larger heaps ventilation passages must be provided.

The Indore method

The compost may be made in pits, bins, or heaps in the open. In high-rainfall areas, bins on a well-drained site are best; pits may become waterlogged and open heaps are too exposed to cooling winds and beating rain. For garden use, bins about 4 ft. × 4 ft. × 4 ft. are ideal; for a small farm I have found that three pits, 10 ft. × 10 ft. × 4 ft., each holding up to 5 tons, are adequate. Bins should be sited as conveniently as possible to the main source of material, it being assumed that it will be carried by wheelbarrow. They should be sited on bare earth (not concrete) and protected from the prevailing wind. The sides of the bins may be timber, concrete blocks, or bricks, and should have ventilation gaps. In the wetter parts of the country, a movable roof is required. Some sort of ventilation shaft should be provided to lead air to the centre of each heap.

If one has three bins in line, one should start by filling bin 2, followed by bin 1. When bin 2 is ready, it is turned into bin 3. In due course bin 1 is turned into bin 2 and so on. Bins are loaded in successive horizontal layers of greenstuff, FYM, a sprinkling of ground limestone, and a shovelful of topsoil or old compost. The greenstuff layers may be about 8 inches deep and the FYM 2 inches. If the materials are dry, a sprinkling of water should be added with each layer. Any coarse greenstuff such as cabbage roots, hedge trimmings, etc. should be chopped up to avoid large air pockets. If different types of FYM are available (horse, cattle, pig, and chicken dung) the more they are mixed in the heap the better; the same goes for different types of vegetable matter.

In winter, greenstuff, which is relatively rich in nitrogen is seldom available; straw bedding in the FYM must provide the cellulose and breakdown is correspondingly slower. It is then more important than ever to include some topsoil to introduce the essential bacteria.

A well-made heap will heat rapidly to about 66°C in the first week, the temperature then dropping slowly down to about 29°C after three months. Aerobic thermophyllic bacteria thrive between 41°C and 54°C, and it is these that cause the rapid breakdown of cellulose.

After about four to six weeks, when the temperature has dropped to about 41°C, the heap should be turned, when a further surge of bacterial activity and a second heating will take place, causing further breakdown of the vegetable matter.

At this turning, the outside, unrotted material should be placed in the middle of the heap. A second turn may be made (ideally) when the centre of the heap has dropped to about 29°C. Thereafter the heap will 'mature' and fix nitrogen from the atmosphere.

Once this process is complete, the compost is ready for the land, and any further storage will result in some loss, unless it is kept covered and turned from time to time.

All-vegetable compost

Gardeners and allotment holders may not keep animals or be near a convenient source of sewage sludge or other animal waste. An alternative source of nitrogen will be required to ensure a rapid buildup of bacteria and the complete breakdown of cellulose. A chemical breakdown can be achieved by the addition of any readily available nitrogen source such as sulphate of ammonia, or Nitro-chalk, but these soluble salts can inhibit bacterial action and the end-product is not so well balanced and does not produce such good results. Some weeds and herbs rich in nitrogen – notably nettles, fat hen, and comfrey – heat and break down very rapidly without the presence of animal waste, and an effective catalyst may be prepared from herbal extracts or bought ready made. The compost is innoculated with a very dilute solution of these extracts which stimulate nitrogen-fixing bacteria and cause the rapid breakdown of the heap. These extracts (sometimes referred to as 'activators' or 'starters') are not substitutes for the nitrogen-bearing vegetable matter in the compost ingredients.

Green manuring and sheet compost

The ploughing in of a quick-growing crop as 'green manure' (mustard, rye, tares are suitable) is sometimes used to increase fertility and improve soil structure. Unless this is done under ideal conditions the herbage may break down slowly, and while the soil population is engaged in this breakdown there may be a retarding effect on the growth of a subsequent crop; the benefit occurring later. As in a conventional compost heap, the process can be hastened by the addition of animal waste, either by very intensive grazing or by spreading FYM or slurry before ploughing. Nevertheless, green manuring should not be considered as a means of stimulating quick plant growth. This method of making compost in the field is known as sheet composting, a variation of which may be achieved after a cereal crop by chopping the straw and stubble (using a forage harvester with delivery spout removed) behind the combine, spreading slurry, and then rotovating the mixture into the top soil. This method is more suitable for large farms with sophisticated equipment.

Earthworms

The vital role of the burrowing earthworm in building soil structure and fertility has been sadly overlooked on modern farms. Indeed it has been almost *necessary* for the agricultural industry to ignore it in order to develop the modern chemical systems of farming. An acre of average British soil under mixed farming conditions may contain 50,000 burrowing earthworms. A well-run organic farm could hold many times that number. But an intensively worked arable farm growing continuous corn with the use of synthetic fertilizers, herbicides, and pesticides may have none at all.

According to the classification of Michaelson (1900) there are eleven families, 152 genera, and some 1,200 species of earthworm through-out the world. The most important family in the temperate and colder regions of the Northern Hemisphere are the Lumbricidae, consisting of ten genera and more than 40 species and sub-species. Of these *L. terrestris, A. longa,* and *A. nocturna* are deep burrowers and of the greatest significance in maintaining soil structure. These burrowing earthworms literally eat their way through the soil forming deep, vertical, permanent burrows with smooth walls cemented together with mucus secretions. The mineral and organic matter that passes through their guts in the burrowing process is mixed and churned in the gizzard, the calcium content is concentrated by the calciferous gland, and the resulting casts, which are deposited on or very near the surface, contain more micro-organisms, inorganic minerals, and organic matter in a form available to plants, than does the surrounding soil.[7]

Worm casts have been shown to contain enzymes which continue to break down organic matter even after they have been excreted. Not only are worm casts rich in plant nutrients, but are in a crumb form some $\frac{1}{8}$ in. in diameter which can be very efficiently exploited by plant root hairs from all sides. Root hairs are almost $\frac{1}{16}$ in. in length.

An average British pasture with 50,000 earthworms to the acre has been estimated to produce about 10 tons of casts (or $\frac{1}{10}$ in. of topsoil) per annum. In the tropics there is much greater worm activity; in the Nile valley more than 1,000 tons of casts per acre per annum have been reported. The value of this annual increment of rich topsoil can hardly be exaggerated. If the population of burrowing earthworms could be increased five-fold (and there is no reason why it should not), the natural improvement in soil structure and fertility could transform the productivity of our land and save us enormous expense. Given the right conditions of air, moisture, non-acidity, organic matter, temperature, and protection from physical damage, the earthworm population will increase naturally.

If one starts with a podzol soil (acid, anaerobic, and without worms) the increase will be very slow as earthworms cannot migrate quickly. In such cases it would be worth the trouble of introducing worms from elsewhere. If, on the other hand, a small but well-distributed population already exists, it will respond on its own as environmental conditions improve, although even here it might well be sensible to import new stock to speed up the process.

The culture of burrowing earthworms for innoculation of soil is described in great detail by T. J. Barrett in his book *Harnessing the Earthworm*. Once the principles are firmly grasped and the necessary boxes or bins constructed, breeding and innoculating the soil with earthworms is neither difficult nor laborious, but care must be taken to keep the culture beds moist but not waterlogged, and the temperature within bounds, and to ensure a plentiful supply of organic matter to feed the growing population.

The non-burrowing earthworms play a different role in soil fertility. In particular *E. foetida* (the stinking dungworm or brandling) multiplies rapidly in a dungheap or compost pit and plays an important part in breaking down the cellulose once the temperature has dropped to about 32°C. The breakdown of compost can be greatly hastened by the introduction of brandlings at the right stage. Details of practical compost-making with help of worms are given in J. Temple's booklet *Worm Compost*, and J. Minnich's comprehensive treatise *The Earthworm Book*, in which he deals with almost every aspect of breeding and marketing earthworms.

Earthworms have extremely sensitive skins and cannot thrive under acid conditions. In the short term, nitrogenous fertilizers, by increasing grasses and clovers, can result in an increase in the earthworm population, but continuous use of superphosphate or Nitro-chalk over a long period has been shown, in trials at Rothamstead, to reduce numbers in proportion to the amount of nitrogen applied.[7] There is further evidence to indicate that nitrogen applications on or near the surface tend to make earthworms feed and excrete their castings well below the surface.[4] In 1978 I attended the Welsh ploughing championships, held on rich alluvial land in the Wye Valley which had been in continuous cereal production for several years. In 200 acres of fresh plough I saw hardly any earthworms at all.

Most pesticides are lethal to earthworms on or near the surface, and some can kill in very small concentrations to a great depth. Herbicides tend to kill worms, either indirectly by killing the plant life on which they feed, or (in the case of fungicides) directly. This is particularly important in gardens and orchards where so many herbicides and fungicides are used.

Soil health: the paramount aim

By building up a good soil structure and providing conditions where all the natural soil flora and fauna can thrive, it is possible to realize the full potential of any piece of land, and to grow on it vigorous, healthy crops and livestock. In these conditions it should not be necessary to use direct-feed chemical fertilizers, herbicides, or pesticides other than to meet an occasional crisis, probably caused by a slip-up in our husbandry. If we fail to build a healthy, living soil, chemicals will, at best, provide a short-term panacea; in the long term they will 'kill the goose that lays the golden egg'.

5 Field crops

The crops grown on a farm and the reasons for growing them can vary enormously. At one extreme is a horticultural holding, with a high proportion under glass, growing vegetables or flowers for sale in a specialized market; at another is an all-grass farm where livestock are reared or fattened. The aim may be cash or some degree of self-sufficiency. The farmer's delight may be in one or more species of livestock, in arable farming, in gardening, or in bringing all these enterprises together in a mixed farm. The land, climate, and markets will also play their part in dictating which crops are worth growing.

Cash crops are always 'extractive' and tend to lower the overall fertility of the farm. Forage crops for consumption on the farm are not directly extractive (the sale of milk and livestock is) although they may transplant fertility from one field to another. Grass leys, grazing catch crops, and crops grown to plough in as green manure all enhance fertility. Crops may be annual plants (the cereals, mustard, Italian rye-grass), biennials (roots), or perennials (most grasses and clovers).

Let us consider the cropping on a small mixed farm with cattle, sheep, and poultry on which you are aiming, as far as possible, to grow all the fodder for your animals (your main source of income) and to feed your family. As we consider each crop, we must assume that the land and climate make its cultivation viable. You will be anxious to have enough grazing for your livestock in spring, summer, and autumn, to make enough hay or silage for their winter keep, to provide additional cereals and roots to supplement the hay, and to grow enough cereals, potatoes, vegetables, and fruit for your own table. You will also want to preserve some well-structured meadows where stock can run in winter without too much poaching. These requirements are bound to conflict on your limited acres, particularly in spring and early summer when a large proportion of your land is laid up for hay or in arable crops. Your most important task will be to reach the best possible balance of livestock, grazing, and arable to suit your land potential.

Grass

The most important crop is grass: it is also by far the cheapest and least laborious to grow and feed. Most grasses are perennials which last almost indefinitely under the right conditions. Some, however, are annuals (notably Italian rye-grass and annual meadow-grass) which only survive by reseeding themselves.

Grassland is generally classified as 'permanent pasture', 'short ley', or 'long ley'. The term 'permanent' is something of a misnomer, for the content of any pasture is constantly changing as conditions become more favourable for one species and less for another; it applies only to the fact that a pasture remains in grass for an indefinite period.

Occasionally one hears of a 'permanent ley', but this is a complete contradiction in terms. A ley, by definition, is a temporary sward of chosen species of grasses, clovers, and herbs which may be designed for a short or long term. If it is left indefinitely, it will revert to poorer, indigenous species which are unlikely to be the most productive or nutritious.

All permanent pasture and most leys contain a large number of grasses, clovers, weeds, and herbs. Some short leys, mostly sown as fertility-breaks in arable rotation or for silage, contain only one or two species of grass or clover. Each species has its own growth habit and ideal environment; and will thrive in some conditions and fail in others. Grazing by different animals, mowing for hay or silage, treading by stock and machinery each favours some plants at the expense of others. Good management, based on a full understanding of these factors, is therefore every bit as important as cultivation and choice of seeds in maintaining economic and nutritious swards. Such good management involves proper drainage, maintenance of a neutral pH, a balance between grazing and cutting, mixed stocking, controlled grazing and resting, the avoidance of physical damage (particularly in bad weather), and the return of organic matter to the soil to make up for hay or silage removed.

Permanent pasture

> To break a pasture makes a man; to
> make a pasture breaks a man.

The old saying points both to the great latent – and often unrealized – fertility stored in the root system of an old permanent sward, and to the difficulty of rebuilding this treasure should it be dug up and spent. In good soil conditions, an old sward properly managed will develop a relatively stable composition, a strong soil structure, a prolific population of fauna and flora, and good natural drainage. It will withstand considerable abuse by livestock and machinery, and, although not as productive in the short term as cultivated leys, nor as early growing in spring or as late in the autumn, it provides much-needed living ground for animals through the winter.

As mentioned earlier, most permanent pastures can be improved in both composition and productivity by good management. When one considers the advantages of permanent pasture, and compares relative costs of improving old swards and reseeding to leys, it will be seen that the apparent increase in productivity of leys may be largely offset by these other factors. Unless it is very well managed, a ley will soon tumble down to indigenous species on a poor soil structure, and the end result will be worse than the beginning. New leys are usually very productive in the first two or three years. They then tend to slump rapidly until about the eighth year (known as 'the years of depression') when the sward is considerably less productive than the original permanent pasture. Thereafter, productivity gradually picks up, as soil life and structure are restored, until it returns to its original permanent status. Total production in the long term is seldom higher than that of permanent pasture. This fact would appear to reflect the land potential to which I referred in Chapter 4.

One should therefore think very carefully and far ahead before deciding to break up a good permanent pasture: it is a slow, costly, and difficult task to get it back.

An important factor in achieving high productivity in grassland is the availability of nitrogen to the plant roots. This is generally supplied by top dressing with direct-feed artificials, and there is no doubt that with the right moisture and soil temperature, earlier and more prolific growth can thus be obtained.

A less direct method is to encourage the growth of clovers in the sward. These leguminous plants have the capacity to fix nitrogen from the atmosphere through their specialized root nodules. Clovers are encouraged by a neutral pH (through periodic liming), by a correct phosphate balance (through applications of basic slag), and by plenty of sunlight. Clovers, being of prostrate habit, tend to thrive in close-cropped swards. Close grazing encourages clovers; repeated hay crops year after year depresses them.

The inclusion of wild white clover, which is generally present in permanent pastures but does not thrive in a phosphate-deficient soil, is particularly important, and there is an old saying: 'There must be clover seed in those slag bags.'

Both grass and clovers recover and grow more quickly when

grazed hard for a short period and then rested for two to three weeks than they do under continuous nibbling.

Short leys

Short leys, down for one or two years, are generally sown as a fertility-building break in an arable rotation; the grazing, hay, or silage taken during the ley being of secondary – albeit significant – importance. The improvement to soil structure through the addition of organic matter from the grass roots and the droppings of grazing animals, together with the corresponding buildup of soil life, reverses the extractive effect of arable cropping.

Short leys frequently consist of Italian rye-grass with or without clovers. Through heavy dressings of nitrogen these produce bulky crops for grazing, silage, and eventually for ploughing in. This system has inherent dangers, as forced rye-grass can grow too quickly to take up important minerals, particularly magnesium, and overdoses of some elements can lock up others in a form not available to plants. Such sward, when grazed, can cause hypomagnasaemia in cattle and sheep with disastrous results.

Medium and long leys

Medium and long leys are put down with the object of increasing grassland production, extending the growing season and improving the nutritional quality of the herbage. They may be designed primarily to be cut for hay or silage, for grazing, or for a combination of cutting and grazing.

Medium leys, down from three to five years, are rather like arable crops in that maximum yield during those years is generally the main objective. The field is then ploughed up, fertilized, and resown to another crop. The maintenance of soil structure and fertility can become a major problem, particularly when the emphasis is on cutting for hay or silage, both of which are 'extractive' crops. The grasses selected for medium leys are predominantly Italian and perennial rye-grasses with small amounts of timothy, cocksfoot and meadow fescue. The rye-grasses, particularly Italian, are very leafy and productive, but tend to be superseded by less prolific indigenous species. Red clovers, alsikes, and trefoil, also very productive but short lived, make up the leguminous portion of these mixtures.

Long leys (often misnamed 'permanent leys') are designed for the long-term improvement of grassland. It is always hoped that the superior species of grasses and clovers will flourish and last for ever, giving an earlier spring bite, more bulk for grazing or mowing, better nutritional quality, and later grazing in the autumn. The extent to which this hope is fulfilled depends upon the land potential, and, within that limit, on management. If the farmer could eliminate all other factors and concentrate only on managing his leys (and the soil on which they grow) he might succeed in minimizing the effect of the 'years of depression' and maintain the new grasses and clovers for a very long time. But one cannot separate farming into tidy compartments: the demands of the market, livestock numbers, weather conditions, and many other variables force one to compromise. The precious, vulnerable new ley gets neglected or misused and gradually tumbles down to a less demanding, less productive permanent pasture.

The grasses for long leys should have very little Italian (annual) rye-grass, as this is liable to die out within two years leaving gaps where indigenous grasses such as *poa annua* will take over. A small amount of Italian may be included to give bulk in the first year. The main grasses are perennial rye-grasses (for leafy bulk and early growth) with a higher proportion of timothy, cocksfoot, and fescues. The clovers should be predominantly wild white which, though less bulky, are hardier and longer-lasting than red clover and alsike.

An important feature of long leys should be their deep-rooting ability. As long ago as the late nineteenth century, Robert Elliot in co-operation with James Hunter, the seedsman, had developed his Clifton Park System of farming, in which the deep-rooting, herbal ley was the cornerstone on which soil structure and fertility were built. He added chicory and other herbs to his seed mixtures, which, through their deep roots, provided communication for water, air, plant root growth, minerals, and nutrients between the living topsoil and the lower strata. Today, mixed herbs, including chicory, yarrow, burnet, rib grass, and sheep's parsley are becoming widely used in long ley mixtures.

Importance of management

The good management of grassland, be it permanent pasture or ley, offers probably the greatest scope for successful farming in the wetter parts of Britain. But the mere fact that grass (of a sort) will take over and grow without effort on the farmer's part, provides a great temptation to exploit this gift of nature without considering the implications. Good grass requires great care and understanding, as described above, and one should never forget the adage: 'To make a pasture breaks a man.'

After ploughing and reseeding, it takes some years for the soil to recover its stable structure, and the grass roots to develop their full strength. During this time overgrazing and treading by stock, particularly in wet conditions, will damage the root system, break down soil structure, and cause gaps in the new sward. These gaps will be filled by indigenous species such as *poa annua* which can form seed at any time during the year. When attempting to establish a long ley, the greatest care should therefore be taken to limit grazing and treading to times of good weather and soil conditions. Extractive hay or silage crops should not be taken in the first year, although the ley may be topped if necessary.

Grass and clover seeds need fairly warm conditions for germination and are more sensitive to drought or excessive moisture than cereals. Thus herbage seeds cannot be sown as early or as late in the year as the cereals so frequently used as cover or nurse crops.

It is only possible to lay down general rules, perhaps the most important of which is that grass, and more especially clover, must always be sown early enough in the growing season to have become fairly established before the continuous wet conditions of winter set in. Thus although it may be quite safe to sow grasses and clovers on dry soils in the regions of low rainfall in the South and East of England in August, and in exceptional years even later, in regions of high rainfall such as prevail in Wales and most of the North and West of England it is likely to be somewhat risky to defer sowing later than about the middle of June.

Although excellent results may follow from sowings in March, yet in many districts the soil is then likely to be too wet and cold, particularly for the smaller seeded species like timothy, rough-stalked meadow-grass, white and alsike clovers to establish themselves satisfactorily. Except in regions of very low rainfall April should be regarded as the safest month for sowing herbage seeds.[8]

I have sown grass seeds successfully by two methods in the high-rainfall conditions of Wales. The first is under a spring-sown crop of dredge corn, usually in April. The ground, previously in roots, is ploughed as early as possible in February or March and the cereal crop sown as soon as a reasonable tilth is obtained. In April, when the corn crop is showing, the ground is lightly harrowed and the seeds broadcast by fiddle or barrow. They are then lightly pressed in with a Cambridge roller. Stapledon strongly recommends a second harrowing to cover the seeds before rolling. When the cereal crop is harvested in August, the seeds are quite well established and capable of standing a light grazing in September and October. The second method is to broadcast the seeds together with not more than 1 kg. to the acre of rape and turnips in late June or early July. The rape should be grazed off in September and October (before the ground becomes too wet) and then the crop rested until the spring. I have found that later sowing of grass by this method is very risky and that too heavy a crop of rape smothers the seeds. It is always wise to remember that the seeds are more important than the nurse crop.

It is necessary to have a clean seed-bed, well consolidated and with soil in good heart. If the soil is acid or lacking in phosphate, lime or basic slag should be applied to ensure a good establishment of clovers. Nitrogenous fertilizers should not be used on new leys (with the possible exception of one-year leys grown for hay or silage).

Grazing systems

The simplest way to feed grass is to allow livestock to eat it where it grows, green and fresh. The further one can extend the grazing season in spring and autumn, without damage to soil or sward, the better. Grazing has the added advantage of returning animal manure direct to the soil. The importance of grazing both cows and sheep over the same land (either together or in succession), both for the good of the sward and the health of the livestock, is explained at the beginning of Chapter 6.

'Set stocking', in which animals are given free range over a relatively large area and not moved regularly, is the most labour-saving system of grazing. The provision of water may well be easier and shelter in foul weather more readily available somewhere in a large area than in a small paddock or strip. The main disadvantages of set stocking are the buildup of intestinal worms and the growth of tussocks of coarse, unpalatable grass where animals have dunged.

'Paddock grazing', in which the grassland is divided into small paddocks, grazed hard for a short period and then rested for up to three weeks, is a very good system for the control of parasite worms, the life cycle of which is broken during the rest periods. If the paddocks can be topped and harrowed after each grazing, a more even, tussock-free sward can be maintained. By arranging 'forward (or sideways) creep' gates between paddocks, calves or lambs can graze the best, cleanest grass ahead of their dams. Water must, of course, be provided in all paddocks.

'Strip grazing' behind an electric fence permits the farmer to ration his grass at a fixed rate. If a 'back fence' is used, the stock is confined to its daily strip and the remainder of the grass is completely rested for a long period. This system is laborious, and if the fencing equipment is not carefully maintained, stock can break out with disastrous results. Single-strand fences will restrain cows; two or three are required for calves, and electric netting or a multi-strand fence is needed to contain sheep. Strip grazing is used widely for cattle on kale, but is less popular with mixed stock or sheep on grass. Water must always be available.

'Foggage' grazing in late autumn and winter, on the hay aftermath which has been allowed to grow tall in late summer, is still practised in some areas, but is a rather wasteful use of grass. Grasses have a tendency to 'winter burn' more or less in proportion to the amount of herbage standing in a field at the onset of frosty weather. It is therefore unwise to allow grass to grow up and mature in late summer.

'Zero grazing', where grass is cut and carried to stock in sheds or yards, is practised in commercial 'feed lots' and on small mixed farms in parts of Southern Europe. This is probably the most laborious system of feeding fresh grass, but it does eliminate the uptake of intestinal worms and saves the fields from poaching. A further disadvantage is the work involved in returning the manure to the field. Nevertheless, I have found that limited zero grazing is a very useful way to feed grass from the orchard and garden to my house cow.

Preservation for winter feeding

Grazing, in any meaningful sense, is limited to the growing season – at best about eight months of the year. For the remaining four months (and in some areas this may be up to seven) grass can be fed only as hay, silage, or artificially dried grass. It follows that a significant proportion of the annual grass crop must be preserved for winter feeding.

Hay

The ultimate value of the hay harvest depends on many factors. Soil fertility, seeds mixture, age of ley, rainfall, sunlight, wind, growth time, stage of growth when cut, harvesting methods, and storage: each can make or mar the final product. The ideal hayfield might be grazed and mowed in alternate years, well drained, not sour (of neutral pH), carrying a well-balanced mixture of grasses, clovers, and deep-rooting herbs. It could be dressed with compost or FYM and shut up to grow from early April (not always possible on a well-stocked farm in a cold spring!). Given good weather it could be ready to mow by late June with a lush green butt, with heads formed but not yet ripe.

Although modern haymaking involves expensive, sophisticated machinery, it is still possible for a smallholder to make and store excellent hay with very modest tackle, and it is hay on which he is most likely to depend for the bulk of his winter keep. It follows that no effort should be spared to ensure a bountiful harvest of top-quality hay, even when this may involve a clash with other priorities.

'Meadow hay' is made from permanent pasture containing a large proportion of indigenous meadow-grasses, clovers, and herbs (some call them weeds). These meadows are slower growing than new leys, and the herbage is generally finer and softer. Good meadow hay is ideal for sheep, who tend to waste the coarser fibres in 'seeds' hay.

'Seeds hay' is produced from leys in which the herbage is predominantly introduced species including rye-grasses, alsike, and red clovers. These mixtures are earlier growing and more prolific than meadow-grass, and can be more difficult to make into top-quality hay.

The harvesting of hay (and cereals) will be considered later; it will suffice to say here that baling is not essential on a smallholding, and if required can be done by a contractor. More storage space is needed for loose hay than for bales, but loose hay is not so likely to heat dangerously. Bales are more convenient for feeding, but with an old-fashioned long-tined hay fork, it is not difficult to carry a good load.

Barn-drying is a safe method of ensuring high-quality hay, and is to be recommended in the wetter parts of the country. An electrically or tractor-belt-driven fan is used to blow air up through the bales or loose hay, which can be baled and carted quite green and at least a day sooner than normal hay. A self-reliant smallholder with a good workshop could set himself up for barn-drying quite cheaply, connecting a second-hand electric or internal combustion motor to a suitable turbo-fan, and leading the air through ducts on the barn floor. In dry weather (when the relative humidity is low), cold air is sufficient, but in close, moist conditions, some heating of the air is necessary.

The rising cost of power for heating and blowing air now makes barn-drying less attractive when compared with conventional haymaking.

Grass silage

Silage-making is a more complex operation than haymaking. It involves a clear understanding of the principles of aerobic and anaerobic respiration, the catalytic actions of enzymes, the control of the balance between the various organic acids produced by respiration, the development of micro-organisms (bacteria and fungi), and the role all these play in decomposition and preservation of the silage crop. A proper understanding of these principles enables the farmer to control the amount of air (and thus the temperature) in his clamp, thereby promoting the correct balance between the desirable lactic and other organic acids. The principles underlying the biochemical changes in the ensiling process are clearly described by Watson and Smith in their book, *Silage*.

Modern silage-making on a large scale involves the use of sophisticated, expensive machinery which is beyond the economy of a smallholding and not suited to working in small fields. It is quite possible, however, to make good silage with the simplest tools. The crop can be cut by scythe, horse mower, binder, or normal tractor mower. It can be put into windrows with hand rake, side delivery rake, or horse rake; collected and carted using wagon, trailer, or buckrake; and made in a tower, clamp, or pit. The main problems in making and storing silage are the exclusion of air during fermentation and holding the clamp together under the pressure that mounts up while it is being built. Strong side walls of earth, concrete, or sleepers are all that are needed in the way of building. Silage can also be made in large plastic bags which must be carefully sealed so that no air can enter. Stacks are not very suitable for small quantities, as too great an area is exposed to the air and a relatively high proportion of the crop wasted. The chief drawback to small-scale silage-making is the extra physical work involved in handling the green crop with its high water content. The practical making of silage using cheap and simple equipment is well described by Watson and Smith, and, if properly made, it is every bit as nutritious and more succulent than good hay.

The location of silage tower, clamp, or pit is important with relation to feeding: it can be a great disadvantage to have to cart silage some distance to the feeding-point. On well-equipped large-scale holdings, self-feed clamps with concrete standings are generally sited close to the cattle-yards. On a smallholding it may be more practical to carry silage to the stock in yard or cowshed.

Before embarking on silage, a smallholder would do well to master the principles and work out clearly in his mind how he is going to cut, gather, and cart the crop, where he will locate his tower or clamp, and how the silage will be fed. Although the crop is put green into the clamp, it does not follow that good silage is easier to make than good hay.

Cereals

Cereal crops – wheat, barley, oats, rye, maize, or 'dredge' mixtures – may be grown for a variety of reasons: for the sale of grain for milling, malting, or animal feed; for home bread-making; for feeding one's own livestock either threshed and rolled or on the straw; for grazing as a green crop; for arable silage; for ploughing in to increase fertility; or for the sale of thatching straw.

Even in ideal cereal-growing country the sale of grain in the open market for milling or feeding is unlikely to be worthwhile for a smallholder who could not produce enough to compete on an adequate scale. In the right location, however, he might find a very profitable market for hand-harvested and threshed wheat and rye straw for thatching and horse collars. It should not be forgotten that *any* crops sold off the land are 'extractive' and tend to lower fertility.

Growing season

To grow grain even for home use, one needs fertile land that is ploughable and a growing season long enough to ensure ripening of the crops. Although oats, barley, rye, and even wheat have been grown (sometimes successfully) in cold, wet areas and at high altitudes, the chances of a good harvest are very much reduced in such conditions. The effective growing season can be extended by autumn sowing with varieties that are winter-hardy. Such crops get away earlier in the spring and are more likely to ripen before the autumn rains. Oats and rye are less 'hungry' than barley and wheat, and will ripen in a shorter, colder, wetter season. Oats are traditionally grown in Scotland and Wales; rye is the staple grain of the cold North European plain and the higher alpine holdings, and could be a very useful bread or animal-feeding grain on a smallholding.

Crop rotations

Threshing crops are always 'extractive' even when the grain is eaten on the farm. It is therefore important to refrain from continuous cereal-growing on the same land, and to put back the equivalent of what has been taken out. This can be achieved, together with crop health and weed control, by adopting a carefully planned rotation of cereals, roots, and grass.

In the best cereal-growing areas, arable farming may dominate the scene, with animals and grass taking second place. There, four- and even five-year rotations, with only short grass leys between to restore soil structure and fertility, are quite normal on the ploughable land; permanent grass or long leys are confined to the steeper or more difficult fields. In the cold, wet lands of the North and West, a much smaller proportion of the land is in arable, the rotations of cereals and roots are shorter with much longer grass breaks between. In the steepest, wettest areas, stock-rearing takes complete precedence, and little, if any, arable farming is undertaken. Between these extremes are found the typical mixed farms with livestock, grass, and arable nicely balanced. The traditional Norfolk four-course rotation of turnips or swedes to be grazed off by sheep in winter, spring barley for sale (undersown), red clover ley to be grazed in spring and summer, followed by winter wheat for sale, was designed to suit the strong land and dry, warm climate of East Anglia. In recent years it has been modified to reduce labour costs and increase profits through the use of machinery, fertilizers, and chemicals. Up to six years of cereal cash crops with a single 'root break' in the middle, followed by a two- or three-year ley is now typical: animals have practically disappeared from arable farms. Some farmers even go so far as to grow continuous cereal crops indefinitely – barley on the lighter land and winter wheat on the strong clays. In recent years the seedsmen and the chemical companies (the latter are rapidly taking over the former) have been producing new varieties of hybrid seeds, fertilizers, herbicides, pesticides, and fungicides, which together have made continuous monocropping feasible. But this system has run into severe problems of falling yields, weeds, and plant disease.

The original Norfolk rotation was used to grow cash crops on strong land, yet never were two successive cash crops taken. Every other crop was grazed off and large quantities of well-rotted farmyard manure were returned before cereals were sown. This system exploited good land to its full potential. The later rotations, if they can be truly so called, burned up the fertility capital at a rate well above true potential. The former system could still be used by a smallholder on suitable land, the latter would be disastrous.

In the colder and wetter areas, much less land is in arable, and less extractive rotations have been developed. Typical of these is the Welsh three-course rotation: oats, either autumn or spring sown; followed by mangolds or swedes; and then barley undersown to a long ley. All these crops are for feeding the farm stock, the swedes often being grazed *in situ* during the winter. The barley and oat straw provide the main winter diet for beef cattle; the rolled oats and rolled or ground barley are fed to poultry, pigs, and bovine youngstock.

There are numerous variations on these basic themes to suit soil, climate, and the main farming enterprises. I find that autumn-sown oats or rye ripen early and often allow me to squeeze in a catch crop of oats or rye after the main harvest to be grazed in the late autumn and spring before ploughing for roots. But in a cold, late spring I find it difficult if not impossible to work up a good tilth in time for mangolds. After roots (and a winter fallow), instead of barley I prefer to grow a dredge mixture of oats, barley, rye, and peas undersown to a herbal ley to be fed on the straw to young cattle in early winter. If I had kept a sizeable pig unit, I would have grown barley for grinding into meal.

When planning a rotation, it is very helpful to draw up a long-term scheme and dovetail the various crops into the seasons, at the same time ensuring that each field comes into the rotation in its turn and is then put back in grass to build up fertility for the requisite period. By this means it is also possible to plan in advance approximately how much of each crop will be grown to meet your needs.

On my own ten-acre farm I grow one quarter of an acre of winter wheat and one quarter of winter rye (mainly for bread), and half an acre of undersown spring dredge mixture (oats, barley, rye, and peas) for feeding 'on the straw'. In our cold, wet climate and on the silty red sandstone soil of South Powys, the rye yields and ripens well. The dredge (mainly oats) grows well to be cut green and made like hay in August. The wheat is less reliable, both in yield and quality, being much slower to get away in a cold spring and less certain to ripen and harden into high-quality baking grain. All the dredge straw is eaten avidly by cattle; the rye and wheat straw are less palatable, but are eaten more readily when put through the chaff-cutter and mixed with chopped mangolds or swedes.

My land is heavily stocked, and one acre of cereals is not enough to provide winter keep for cattle, sheep, poultry, and horses. But I could not grow more grain without causing an even worse grazing crisis from April to July when three acres are laid up for hay. The value of a long-term rotation plan in solving this equation can hardly be overstressed (see Figure 3, p. 46).

Wheat

Wheat is a hungry crop, needing strong, neutral soil and a long, warm growing season. Originally a native of Mediterranean lands, it has never adapted completely to cold, wet conditions.

Wheats may be hard or soft, strong or weak. Hard wheats are easily milled into white flour; soft wheats do not separate so easily. Strong wheats contain a large amount of good-quality gluten providing an elastic dough which rises well for baking bread. Weak wheats do not rise and are suitable for biscuit-making.

Some wheats require a period of cold weather and short days if they are to yield well; these are autumn sown, stand over winter and ripen much earlier than spring-grown varieties.

Recommended varieties

	Winter	Spring
Good bread-baking qualities (strong)	Maris Widgeon	Timmo
	Maris Freeman	Sappo
	Flinor	Sicco
	Flanders	Highbury
Biscuit quality (weak)	Maris Kinsman	Maris Butler
	Maris Ranger	
	Hobbit	

Wheat should be sown early in the rotation when the soil is at its most fertile. Autumn sowing gives the best chance of ripening, particularly in the colder, wetter parts of the country. Autumn-sown wheat (and other cereals) should have a rough seed-bed to avoid soil capping, particularly on silty soils, and to provide some protection to the seedlings from icy winds.

Experiments are being carried out at Cambridge on a wheat–rye hybrid known as Triticale, which, it is hoped will have the growing and ripening qualities of rye and the baking qualities of wheat. This could be a godsend to the smallholder, especially on poor, cold land. I grew Triticale for three years

with moderate success. The growing and ripening characteristics were most promising and the grain made excellent bread.

Emmer bearded wheat, common some 2,000 years ago, is
reputed to have a very high protein content and good baking
qualities. It is extremely difficult to thresh, but is relatively
bird-proof – a significant advantage in areas where very little
grain is grown. Experiments are now being conducted near
Reading to determine yields and to develop a suitable
threshing technique. I have been growing small plots for some
years, but have not yet solved the problem of threshing.

Barley

Barley is grown for malting in the manufacture of beer and
whisky and for animal feed, particularly the fattening of pigs
and intensively reared beef. It is less 'hungry' and earlier
ripening than wheat, and grows well on lighter, well-drained
soils provided they are not acid. When grown on strong, very
fertile land, barley tends to lodge; it is therefore grown in the
later stages of a rotation when the fertility has been reduced.

The main interest of barley to a smallholder is as stock feed,
particularly for fattening pigs or poultry. For this purpose it is
ground into a coarse meal. When used in cattle rations, it is
rolled like oats, but is not, in my opinion, as good as oats for
milking cows, which need more protein and less carbohydrate.
Barley straw is soft and palatable, and is widely used as bulk
feed for beef cattle.

Barley may be autumn or spring sown, slightly shallower
and on a finer seed-bed than wheat.

Recommended varieties for feeding

Winter-sown	Athens, Igor, Sonja
Spring-sown	Athos, Aramis, Magnum, Midas,
	Lofa, Abed, George, Mazurka,
	Maris Minute, Sundance, Goldmaker

If allowed to go overripe, barley ears tend to fall off the stems, and heavy losses can occur. The crop is often cut slightly green to avoid such loss.

Oats

Oats are mainly used for feeding all types of livestock except pigs whose digestive systems cannot cope with the fibrous husks. The best-quality oats are also used for breakfast foods and oatcakes. Oat straw, although variable in quality, generally makes good bulk fodder for cattle.

Oats grow best in the wetter, cooler parts of the country and are much more tolerant of poor, acid soils. Cultivation and sowing are similar to that of wheat and barley; both spring- and autumn-sown crops are common. Winter oats may be grazed in autumn and spring before growing up to head; this practice encourages 'tillering' and higher yields. Like barley, oats are often cut slightly green to avoid shedding of the grain.

The husks of oats are not separated from the grain by normal threshing, and the grain is either rolled or fed whole.

Recommended varieties

Winter-sown	Maris Osprey, Pennal, Peniarth
Spring-sown	Leanda, Mostyn, Margam, Maris Tabard, Maris Oberon, Karin, Ayr Commando, Maelor

Rye

Rye may be grown as a grazing crop or for threshing for human or animal feed. It has never been popular for bread-making in Britain, although it is the staple diet of the cold North European plains and in the Alps. It has long, tough straw which is much sought-after by thatchers and harness-makers, and, if put through a chaff-cutter and mixed with chopped roots, is readily eaten by cattle. Ryes and rye–wheat hybrids are sometimes prone to ergot – a fungus disease that destroys the individual grains and replaces them with black spores which look like mouse droppings. These are poisonous, and any crop showing ergot should not be eaten.

Rye is very tolerant of poor, acid soils, is extremely frost-hardy and ripens much earlier than other cereals. In the cold, wet northern and western parts of the country it is by far the most reliable cereal crop. It threshes easily and rolls like oats to make a palatable stock feed. Rye flour does not rise on its own, but may be mixed with wheat in different proportions and can be baked into a variety of tasty breads.

As rye will grow in very cold conditions, it is probably the best catch crop for autumn and spring grazing. For this purpose it is autumn sown after wheat or barley and before roots. In a cold, late spring it may prove difficult to graze off the rye in time to work up a good tilth for mangolds.

Unlike other cereals, rye is cross-fertilized and does not remain true to type. Fresh seed should therefore be bought each year.

Recommended varieties for autumn sowing

For grain   Otello, Ashill Pearl, Dominant
For grazing   Lovazpatonai, Ovari, Rheidol

Maize

Maize may be grown for human consumption (corn on the cob), animal grain, or arable silage. It is not widely grown in Britain, as it requires strong land and considerably higher temperatures than other cereals. Smallholders in sheltered, low areas in the South-East of Britain, may grow a little maize in the vegetable garden; otherwise the crop is of little interest to them.

Dredge mixtures

Dredge mixtures of cereals, with or without peas or beans, produce heavier yields than when the crops are grown separately. This is partly due to the ability of pulses to fix nitrogen from the atmosphere, which feeds the rest of the mixture, and partly due to the little-understood symbiosis between 'companion plants'.

Although care should be taken to choose varieties that ripen more or less at the same time to avoid losses, dredge mixtures are generally cut green and dried like hay to be fed 'on the straw' or made into silage. They may be sown in spring or autumn and undersown to a ley.

I use a spring-sown mixture of oats (60 per cent), barley (20 per cent), rye (10 per cent), and peas (10 per cent), undersown to a herbal ley as the final crop in my three-course rotation. I cut the crop quite green and dry it on a Tyrolean-type wire fence (see p. 108). I have found the crop to be very reliable in our wet conditions, and it is relished by cattle in early winter. If it is stored for too long in rick or barn, rats and mice can cause heavy losses.

Roots

Root crops – mangolds, swedes, common turnips, rape, and kale – are of the greatest value in the winter feeding of cattle and sheep. They provide a succulent supplement to hay and straw which helps to stimulate and maintain the milk flow through the lean winter months until the spring grass begins to grow.

Mangolds

Mangolds are of the 'beet' family with the characteristic multiple seed. They require a fine tilth with plenty of humus and moisture, and are slow growing. They are normally sown 'on the ridge' as soon as the ground is ready in early May, and hoed and singled at the two- to three-leaf stage whilst it is still easy to separate the entwined stems which grow from the multiple seeds.

Mangolds must be lifted and clamped before the risk of any hard frost. As they are liable to 'bleed' they should be topped with some stem left on the tuber, and the roots cleaned but *not* trimmed. They are normally left to 'sweat' for ten days in small heaps in the field, covered with the leaves, before being carted to the clamp.

Although mangolds are not frost-hardy, they keep very well in the clamp, and are fed, after maturing, from February to June.

Swedes

Swedes, like rape, cabbages, and turnips, belong to the *Brassica* family and are extremely frost-hardy. The seeds are very small, and should be drilled shallower than mangolds (about one inch), either on the flat or on the ridge, in a well-worked, humus-rich seed-bed. They are hoed between the rows and singled by hand. They can be heavily attacked by turnip flea beatle when in the two-leaf stage if the seeds are not dressed, but will usually outgrow the damage. The brassicas are susceptible to a fungus disease called finger and toe, particularly in acid soil, and, if sown before mid-May, are liable to mildew. They grow best in cool, damp conditions, and will continue growth well into the autumn.

Being frost-hardy, they are often grazed off *in situ* through the winter, although this can cause damage to the soil structure. Alternatively they can be lifted and clamped up to mid-December, and fed to cattle and sheep until the end of February. They do not keep in the clamp quite as well as mangolds. When fed to cattle they should be chopped in a root-pulper and, ideally, mixed with chaffed straw. They are generally fed whole to sheep, although this can cause older ewes to shed their incisor teeth prematurely and thus shorten their useful lives.

Recommended varieties

| Dark purple-skinned | medium to high food value, highly frost-resistant, not a high yielder |
| Green-skinned | high food value, late-maturing, fairly frost-resistant, more resistant to club-root disease, higher yielding |

Kale

Kale is grown for feeding dairy cattle. A member of the *Brassica* family, it requires the same fine, neutral, humus-rich seed-bed as swedes. It may be strip-grazed behind an electric fence, cut and carted to the cattle-yard, or made into arable silage. For the smallholder kale is a rather laborious crop to feed. I have occasionally grown a field of kale and swedes in alternating six-row bouts for strip grazing. The kale (frost-hardy thousand-headed variety) acts as a wind-break to protect the swedes, and is relished by sheep as well as cattle. The thousand-headed and dwarf thousand-headed varieties are the hardiest; the marrow-stemmed are more palatable but not so frost-resistant.

Turnips

Common turnips are quicker growing than mangolds or swedes, but are not frost-resistant and do not keep so well in a clamp. They are generally broadcast with rape in midsummer as a nurse crop for a grass ley, and grazed off before Christmas for fattening lambs.

The white-fleshed varieties are the quicker growing and heaviest croppers, but are the least hardy and poorest keepers.

The yellow-fleshed, 'hardy green' turnips are more frost-resistant and better keepers, but slower to mature.

Forage rape

Rape is an open-leaf brassica which yields a quick, heavy crop of palatable foliage. It is most frequently sown with turnips (see above) as a nurse crop for grass. The most common mistake is to broadcast too much seed producing such a heavy crop that the grasses are smothered. For a nurse crop $\frac{1}{2}$ kg. (or at the most $\frac{3}{4}$ kg. plus $\frac{1}{4}$ kg. of turnips) to the acre is ample; it is not easy to broadcast such a small quantity by itself, but mixed with the grass seeds, it can be sown successfully by fiddle or seed barrow.

Potatoes

Few smallholders will be interested in growing potatoes as a cash crop. In a few upland areas, however, it might be profitable to grow a small patch for certified seed. Most smallholders will grow enough for their own use in the vegetable garden or plant a few rows with the swedes or mangolds in their roots field.

Potatoes need soil with plenty of humus, which should be worked into a good fine tilth. The ground is normally ridged, the tubers are planted in the furrows and the ridges are then 'split' to cover them. The smallholder, growing only a few rows, will probably 'dibble' the tubers into the original ridges by hand, thereby saving the subsequent ridging operation.

If possible the crop should be lifted when the ground is reasonably dry, the potatoes are then spread on a dry barn floor before sorting into 'ware', 'seed', and 'chat' sizes – the latter for animal feeding. They should be clamped in reasonably dry conditions with adequate protection against frost. I store potatoes in sacks under straw in a loft. A note of warning: do not be tempted to wash the mud off potatoes with a hose, as they will not store well in the clamp if you do.

Other root crops

There are several other crops which may be of use to you to suit particular conditions. If you are near the right markets, you may concentrate on horticultural crops for sale, but that is beyond the scope of this book.

Cabbages and other brassicas may be sown in specially prepared seed-beds and transplanted into the fields, or may be sown like swedes direct and singled later. The 'headlands', where the horses or tractors turn at either end of the ridges, constitute too large a proportion of a smallholder's roots field to be allowed to lie fallow, but they cannot be cultivated and sown until the inter-row hoeing is completed. To overcome this waste, I sow several hundred extra brassicas in the greenhouse in March (cabbage, red cabbage, brussels sprouts, cauliflower, and broccoli), bed them out at four-inch spacing in the vegetable garden in early May and then transplant them at two-foot intervals into the headlands after horse-hoeing in June. Waste of land is thus avoided, and there is a surplus of top-quality produce for sale in late autumn and winter. Any unsold can be fed to cattle or sheep.

Cow cabbage is a useful, frost-hardy crop that can yield as much or more fodder than swedes. It can be grazed *in situ* or cut and carted. It does not store in the clamp.

Catch crops

Catch crops, which are grown as a bonus between the main courses of a rotation, can be very worth while provided they do not interfere with the cultivation and growing of main crops, cause undue proliferation of weeds, or form a 'disease link' between susceptible main crops.

Catch crops are usually sown in late summer or early autumn immediately after harvest, for grazing in early winter and again in the spring before ploughing for the next main crop. They may also be grown to plough in as green manure.

Forage catch crops include rye, oats, rape, turnips, and Italian (annual) rye-grass. Mustard, tares, and vetches are very quick growing and make good green manure; mustard is also useful as a tonic when grazed by sheep before ploughing in.

Catch crops should be sown after minimal cultivation of the stubble as soon as the harvest is cleared. In the North and West it may not be worth while sowing a catch crop after a late harvest, particularly if the ground is wet. In a late spring, the catch crop may not be ready to graze off in time to prepare a good tilth for the ensuing main crop.

Field cultivations

Cultivations are carried out for three reasons: to improve the soil crumb structure and so create an ideal seed-bed for the crop; to uproot and kill weeds; and to bury turf or crop residues where they will rot down and not compete with the ensuing crop. The methods used vary enormously with the type and depth of soil, the climate, and the crop to be grown.

Every field operation using tractors or horses inevitably causes some compaction and damage to soil structure; it should therefore be the farmer's aim to create the desired effect with the minimum of passes at exactly the right time. This is an art which can be mastered only with experience.

Every operation will cause the loss of some soil moisture; in dry conditions, therefore, cultivation should be kept to a minimum. The danger of working wet land has already been mentioned in Chapter 4.

Mechanical tillage (apart from rolling) is designed to turn over soil or break down clods into tilth crumbs of a suitable size. Overdone, it can create an unstable, flour-like tilth which will turn to porridge and pan in wet weather, preventing water from percolating to the subsoil. Carried out in the wrong weather conditions it can produce the opposite to the desired effect. The ideal structure is created by plant and animal life in the soil: man's efforts to produce it mechanically are, at best, a poor substitute for nature. Nevertheless, where land is cropped regularly, there are bound to be times when the soil life is reduced or inhibited to a point where it fails to rebuild the damaged structure. Cultivations are then the only alternative until a grass break or green manure can allow soil life to regenerate and resume its proper role.

Ploughs

To bury turf or stubble, one must turn the topsoil over, either with a spade or a plough. The living topsoil is seldom more than six inches deep; if one digs or ploughs deeper, one inevitably puts subsoil on the surface and buries the active flora and fauna below the level where they can thrive and do most good. For burying, a long-boarded general-purpose or lea plough does the cleanest job, turning up an unbroken, shiny furrow.

To break up clods, aerate the topsoil, and create the beginnings of a crumb structure, a semi-digger or digger plough is preferable. It has a shorter, more abrupt mould-board, which breaks the furrow but does not bury rubbish so cleanly.

Ridging ploughs, with double (left and right) mould-boards, are used to set up ridges for potatoes and roots. They vary in size from hand-operated garden tools to three-bodied tractor-mounted implements.

Cultivators

For stirring the topsoil without inverting it, and for breaking down half-rotted turf and clods, a wide range of cultivators is available.

Disc harrows consisting of gangs of concave steel discs are dragged at an angle to the line of draught. They are used to cut up turf and consolidate a seed-bed, particularly after ploughing an old sward. Some highly skilled farmers, particularly in dry areas, prefer disc harrows to the plough, believing that it is never wise to invert the topsoil. In wet areas, weeds and turf re-root and grow quickly and it is not easy to work up a clean tilth unless they are well buried and allowed to rot.

Chisel ploughs are heavy rigid- or spring-tined cultivators which break up and stir the topsoil. They can work at speed, and make several passes in less time than it would take to plough the same area with mould-board equipment. Like discs, they do a cleaner job in dry than in wet conditions. They require a powerful tractor, and are unlikely to be used on a smallholding. Nevertheless, the principle of chisel ploughing has much to recommend it in the right conditions.

Rigid-tined cultivators or 'grubbers' are very useful for breaking up consolidated or panned soil (for instance, in the spring after a root crop), and can often replace the mould-board plough in the first stage of working down a tilth. They require considerable power, and are easily damaged on stony ground.

Spring-tined cultivators with tines at about six-inch spacing, are much lighter than chisel ploughs. They can work down an excellent tilth after ploughing so long as there is no unrotted turf. They do not readily penetrate a panned surface.

Rotovators and other cultivators driven by the tractor power take-off can produce a fine tilth in one pass, and are sometimes used to replace all other implements. They are particularly useful when preparing a seed-bed for roots, but can destroy soil crumb structure by reducing it to 'flour' causing desiccation in dry weather and panning in wet. They also have a tendency to chop up the roots of couch grass and docks, causing the spread of these weeds. Tractor-mounted rotovators are expensive to buy and to operate, but small, manually controlled machines are useful for garden and inter-row field cultivations on a smallholding.

Spike harrows come in many shapes and sizes; the zig-zag, with staggered rows of spikes, being the most common. They are used for the final stages when working down a tilth and levelling the surface once the large clods and turf have been broken. Heavy spikes can be useful in breaking a surface pan in grassland or in growing cereals, and for aerating and reducing moss in swards. I use light spike harrows for covering cereal seeds sown direct onto freshly ploughed land.

Chain harrows generally have spikes on one side and are plain on the other. They are used to cover grass seeds and for final levelling of a tilth, but their main use is for spreading dung and molehills and scratching the surface of grassland after grazing or before shutting up for hay.

Hoes

Hoes are used for the control of weeds between the rows of root crops. Various blades or discs can be fitted to uproot the weeds either between ridges or on the flat. Great care must be taken to set the hoe up and steer it accurately so that the blades run close to the crop without damaging the seedlings. Accurate ridging and drilling make hoeing much easier. Inter-row hoes may be single-row (horse-drawn) or multi-row tractor-mounted implements.

Hand hoes of the Dutch or triangular pattern are used for singling roots and weeding in the row. This task is straightforward in a fine, well-worked tilth, but can be a nightmare in badly cultivated soil containing unrotted turf, and if inter-row hoeing has been inaccurate.

Rollers

Rollers are used for consolidation of the surface to conserve moisture and press seeds into contact with the soil. They are also used, on occasions, to break large clods. They should be used with care to avoid over-consolidation and destruction of soil structure, and never used in wet conditions. Flat rolls are used for levelling grassland and burying small stones before shutting up for hay. Cambridge rolls, with independent ribbed rings, are ideal for pressing cereal and grass seeds into the soil. The ridged surface that results tends to prevent panning.

*

To acquire the art of field cultivation is a lifetime's work. You must know your soil intimately in every corner of every field and at every stage of its structure through the seasons. You must understand the climate and how it affects the soil. You must master the characteristics of your implements and power sources until you can handle them instinctively in all conditions, and you must keep them well maintained so that they are always ready to be used when the right moment occurs.

Direct drilling

Direct drilling is a new technique in which seeds are drilled direct into the ground without previous cultivation. Special drills have been developed with heavy discs or modified rotovators instead of the usual light coulters. These cut slits in the turf or stubble into which the seeds and fertilizers are dropped. There are enormous advantages to this system: the saving in time, the preservation of soil structure, the conservation of moisture, the buildup of organic matter in the topsoil and the maintenance of the earthworm population (normally reduced by conventional cultivations). There are also disadvantages which could, on balance, rule the system out for organic farmers and smallholders: turf or trash are not buried, so stubble must be burnt or grass destroyed by a herbicide before drilling takes place. It is claimed that the herbicides used (including paraquat, delapon, aminotrazole, and glyphosate) are neutralized on contact with the soil and do not harm soil life, and that earthworms *actually increase* under this system. If this could be proved *beyond doubt*, direct drilling would fit in well with the organic philosophy.

The drills are too expensive for a smallholding, and even in the hands of a contractor, the technique (including spraying) is not cheaper than conventional tillage. Perennial weeds such as couch or coltsfoot are more difficult to control, and slugs, which shelter and multiply in the deep, hard-sided slits, can become a serious pest. Winter cereals are slow to establish, and must be sown earlier: this is apparently because there is a tendency for stubble or killed turf to be pushed down into the slits with the seed where it can create anaerobic conditions leading to the formation of toxic substances such as acetic acid. But one might be forgiven for doubting if this is the most likely cause of the toxic side-effects of direct drilling!

Harvesting hay and cereals

The value of all crops, but particularly hay and cereals, can be greatly reduced by bad luck or bad management at harvest time. The weather can upset the best laid plans, but the man who is ahead on his work and has his equipment ready in good order can take quick advantage of short spells of good weather and keep his losses to the minimum. The farmer's prayer is to have good weather when his crop is at just the right stage to reap; the ability to judge the state of a crop can come only with considerable experience.

Hay

A hayfield is normally cut in swathes in a clockwise direction, starting one swathe in from the hedge and finishing in mid-field (see Figure 6). The butt of the first swathe is then raked about 9 inches inwards (towards mid-field) to clear the edge of the final or 'back' swathe. This back swathe is then cut in the opposite direction. This system is used whether the field is cut by scythe, cutter bar, or rotary mower.

After cutting and initial drying of the exposed surface the swathes are turned over by hand or mechanical rake to expose the wet under-surface and to move the whole windrow onto

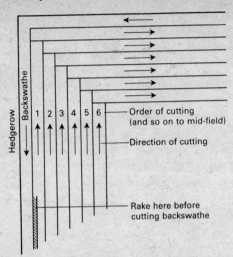

Fig. 6. Mowing a hayfield

dry ground. It is then shaken out with fork or machine to aerate the hay and allow the wind to pass through it. As the edges of the field are normally the most shaded and sheltered from the drying wind, the outside four to six rows are first raked towards mid-field and the rest towards the hedge. This provides one double row four to six rows in from the hedge and this row requires specially careful tedding. The second turning starts at the outside edge turning the whole field including the double row towards the hedgerow. The third turning starts in the resulting gap in mid-field and turns the whole field towards the centre, and so on. Getting the cut hay fit to cart or put on tripods may take anything from two days (in ideal conditions) to several weeks. A typical day's turning and tedding could start as soon as the dew is off the ground. The windrows or swathes are then turned over onto dry ground. This turn is followed immediately by tedder or hay forks and the swathes fluffed up and aerated. If the day is windy without much sun, the swathes should be high and narrow; if very sunny and without wind a wide row is better. Particular attention should be paid to the double row, to hay lying in the shadow of trees and hedges, and to the inevitable lumps at the corners. The hay may then be left until mid-afternoon and the whole process repeated. Before sunset (and particularly if a shower is expected) each row can be raked into a large number of small cocks, or rowed up into narrow swathes. This reduces the surface wetted by dew or shower and exposes a greater area of bare ground to the early sun the following morning. It is then possible to start the next day's turning a bit earlier than would otherwise be possible.

When the hay is fit to cart, bale, or tripod (and only experience can tell you when it is 'singing') it may be raked two rows into one (and possibly four into one). If it is to be baled or carted it is most important that it is well and evenly dried. It is often possible to bale the centre of the field a day before the edges and double row.

Tripodding of hay and cereals is practised widely in the Alps and Scandinavia where good drying conditions are rare. The principle is to support the crop above the ground and to allow the air to circulate freely through it. The crop settles onto the tripod and forms a natural 'thatch', which sheds all rain except in a severe gale. The outside thatch is bleached by sun and weather, but the crop inside dries green and succulent, providing very high feed value.

Hay may also be put into cocks on the ground. If it is already well dried and can be carted in a few days, this method involves less work than tripodding. But cocks cannot stand much bad weather, and the portion on the ground is usually spoiled.

Crops can be safely cocked or loaded onto tripods at least a day before they are fit to bale or cart: a most important factor in unsettled weather conditions. Although tripodding involves an additional operation (and unnecessary extra work in good weather), in high-rainfall areas it will often reduce the need for extra turning and tedding and the risk of crop loss or damage. Once on tripods the crop is safe until help is available and conditions right for carting.

There are many designs of tripod varying from interlocking A-frames to simple uprights made from the tips of larch or spruce trees. Thin crops on exposed hillsides are better on the smaller supports. A good general purpose tripod using the minimum material for the maximum load is illustrated in Figure 7. You will need about ten tripods per acre.

Spruce trimmings from forestry plantations are cheap, and with the heavy wire triangles make a strong and effective structure. These tripods can be dismantled and stored in the minimum space and used without maintenance for many years.

The method of loading (see Figure 8) is simple; you start at one corner of the loading poles and work clockwise around the tripod, keeping the corners well built up. The hay is laid across the poles (which are supported on the trestles) and lower wire triangle, leaving a small air space where the hay meets in the middle of the tripod.

Similar loading continues onto the upper triangle (no space can be left in the middle), still keeping the corners well built up, and finally a cap is built over the tripod tips. The final shape is as shown: the top is narrow and the sides slope steeply. The trestles and loading poles are removed and the entire weight then rests on the tripods and wires. Finally a continuous length of baler twine is threaded under the legs and over the top, pulled quite tight and tied halfway up the load. This can be tightened daily as the load settles, and will protect the stack against wind. Figure 9 shows how to set up a tripod on a slope.

Cereals for threshing

On medium-sized and large farms, cereals are normally harvested by combine, which cuts, threshes, and winnows the

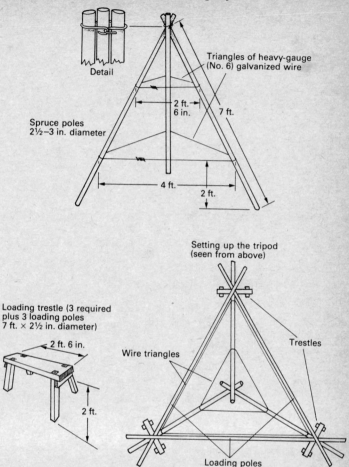

Detail

Triangles of heavy-gauge (No. 6) galvanized wire

Spruce poles 2½–3 in. diameter

2 ft. 6 in.

7 ft.

4 ft.

2 ft.

Setting up the tripod (seen from above)

Loading trestle (3 required plus 3 loading poles 7 ft. × 2½ in. diameter)

2 ft. 6 in.

2 ft.

Wire triangles

Trestles

Loading poles

Fig. 7. The basic tripod with trestle and poles for loading

grain in one operation. The straw (and weed seeds) are left on the field to be burnt or baled later. The crop must be cut before it is dead ripe to avoid shedding of grain, and, because it does not stand in the sheaf to harden, it usually requires artificial drying. A combine, which only works for a few days

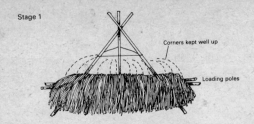

Stage 1

Corners kept well up

Loading poles

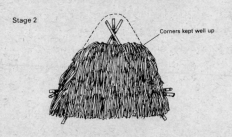

Stage 2

Corners kept well up

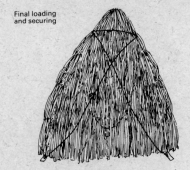

Final loading
and securing

Fig. 8. Loading the tripod

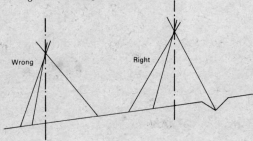

Wrong

Right

Fig. 9. Setting up a tripod on a slope

or weeks in the year is far too costly for a smallholder, who must either rely on a contractor or work with simpler equipment.

The mowing of cereals for threshing involves binding into sheaves with the grain heads at one end. This may be done with a reaper and binder or by scythe fitted with a special cradle. Cutter-bar and rotary mowers are rather unsuitable because they leave the swathe lying in the direction of cut with heads and butts mixed; it is then difficult to pick up, sort, and bind the sheaves. Old binders can still be found at farm sales, but few are in good working order and spare parts and canvases may cost more than the machines themselves.

On a smallholding the scythe is still a sensible, economic proposition in the hands of a skilled workman. A razor-sharp edge is essential; by the time the operator has achieved this, he will also have acquired the basic mowing swing and overcome the inevitable backache. There is no substitute for practice and experience.

The normal swing of a scythe (when mowing hay) is a circular sweep from right to left. This cuts most crop for least work but leaves the windrow in a continuous 'rope' with heads and butts mixed. A modified swing, straight across the line of the windrow leaves the butts to the left but this is harder and slower work. By fitting a light wooden cradle to the scythe (see Figure 10) the normal circular swing leaves the crop lying with butts to the right ready to pick up and bind. The binder can then shuffle down the windrow gathering a sheaf onto his insteps; he then picks it up and binds it with straw or twine (see Figure 11).

The sheaves can then be 'stooked' into fours, sixes, or eights, or loaded onto tripods to dry and harden. In an exceptionally dry year (when straw is already dried out at mowing time) the corn may be carted without stooking, but normally it should be left in the field for a week or more for the straw to dry and the grain to harden. When carted to barn or rick the sheaves

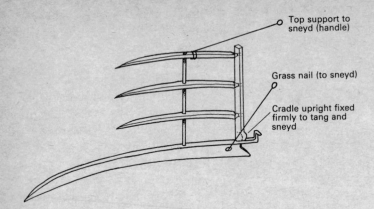

Top support to
sneyd (handle)

Grass nail (to sneyd)

Cradle upright fixed
firmly to tang and
sneyd

Fig. 10.   Scythe cradle

Straw band
(if tied with twine,
a reef knot is best)

Fig. 11.   Sheaf tie

are stacked butts-out in such a way that the heads are not exposed to weather or birds. It is then wise to thresh as soon as possible to avoid loss to rats and mice.

Dredge crops

In wet countries, cereals are often made like hay and fed to livestock 'in the sheaf' (in other words, straw and grain together). Oat and barley straw is softer and more palatable than wheat and rye, and thus the most suitable for feeding in the sheaf. Dredge crops, consisting of mixtures of oats, barley, and peas, with sometimes a little wheat, are also fed in this way. The crop is normally spring sown and should consist of varieties likely to ripen together. The crop is harvested a bit green to avoid the seed being cast, and made like hay. It may be baled or carted loose: baling gives greater protection against rats and mice. Cereal straw is much coarser than hay, and air circulates more easily through it. You can therefore put such a fodder crop onto tripods or a fence when damp, confident that it will dry out, so long as it is not loaded too tight. These crops may be cut with tractor-mower as the lie of the straw is unimportant.

A Tyrolean fence is most suitable for drying dredge crops (see Figure 12). After first construction and adjustment it can be erected and dismantled quickly: provided anchorages and connections are well made, it is trouble-free. It is loaded from the bottom wire upwards and the natural settling of the crop makes big air tunnels under each wire. It is better than tripods for a really wet crop since the crop may be put up the day it is cut even in rainy weather. Fences may also be used for hay, but are not suitable for corn bound in sheaves. You will need about 150 yards of fence per acre.

The foundation of a successful fence is its anchorage. If trees or solid gate posts are not available – and they are seldom found in the right places – strong artificial anchorages must be provided. These may be made from fencing stakes driven at

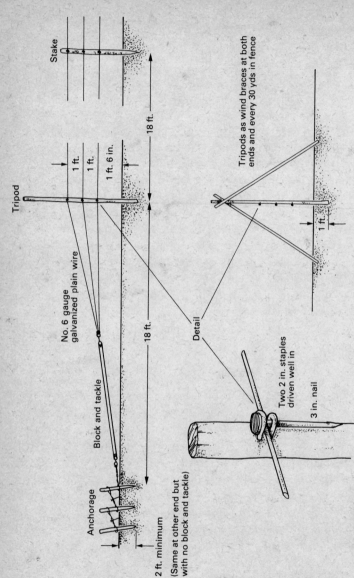

Stake

Tripod

1 ft.
1 ft.
1 ft. 6 in.

18 ft.

No. 6 gauge galvanized plain wire

Block and tackle

18 ft.

Anchorage

2 ft. minimum

(Same at other end but with no block and tackle)

Tripods as wind braces at both ends and every 30 yds in fence

1 ft.

Detail

Two 2 in. staples driven well in

3 in. nail

Fig. 12. Tyrolean fence

an angle, the top of the first being windlassed back to the bottom of the second, and so on. It is well worth spending time to get the anchorages securely placed in line with the fence. It can be infuriating if an anchor gives when the fence is loaded. It is equally important to ensure that all joints and connections between the wires, uprights, and anchorages are properly made and strong enough for the job (see Figure 13).

When first setting up a fence it takes time to adjust the three wires at the tackle end so that they are under equal tension (the far end having already been secured to a strong chain

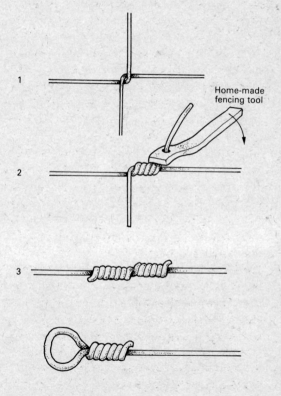

Fig. 13. Making wire joints

link, which in turn is fastened to its anchorage). When approximately equal tension has been achieved, with no load on the fence, the three wires can be marked at the correct length and fastened securely to a similar chain link which will eventually receive the hook of the straining tackle. The fence is then strained fairly tight and is ready to load, the block and tackle being left in place. The weight of the load will tighten the wires and put a tremendous strain on the whole system.

It is equally important to see that the fence is well braced laterally with tripods at the ends and every 30 yards along its length. A gale may otherwise blow the whole fence over. The standard tripod used for haymaking (as earlier described) is ideal for this purpose; the middle leg is dropped into a one-foot-deep hole and the other two spread at right angles to the line of the fence.

The crop is raked towards the fence from either side and loaded onto the wires with a fork, starting with the bottom wire, until there is no room beneath the second wire. It will settle and drape itself on either side forming a natural, water-shedding thatch. The same procedure is followed on the second and top wires, the latter taking a bigger load than the first two. The following day the whole fence can be raked down from both sides to make it shed rain better, and the rakings piled on top.

Before dismantling the fence, it is wise to number the three wires as they are of slightly different lengths. They may then be rolled onto a simple reel, which may be made from a 5-gallon oil drum, and are ready for next year. This type of fence is very quickly and easily erected in the second and subsequent years if the materials are carefully put away; it is the initial preparation and setting up that takes the time.

Harvesting by simple methods, and drying crops on tripods and fences, may seem, at first sight, a wastefully laborious business. At the other extreme, modern, mechanized methods would make economic nonsense on a smallholding. The secret is to find the right balance in which the *safety and high quality of the harvested crop* is of paramount importance. If you can get the crop off the ground and impervious to weather quickly, it hardly matters if the subsequent carting and storing are laborious – wait until you have guests to help. In a good year you may have done unnecessary work; in a bad one the harvest could be safe whilst others are still struggling with their expensive machinery.

Storing hay and straw

It is possible to store hay, straw, and corn in the sheaf in ricks outside. Good drainage and foundations are needed in the rick-yard to keep the crops above the wet, and a skill – rare these days – is needed to build and straw-thatch a stable rick. Feeding is difficult and laborious, and it is not easy to keep the rick waterproof, even with a plastic sheet, as it is used up. Until you have time and money to build a barn, a rick-yard may be the only alternative. If you have no permanent buildings, probably the quickest and cheapest temporary expedient is a plastic sheet with a sisal net to hold it down. A second-hand railway tarpaulin is stronger but more expensive. It is difficult to feed hay from a rick so covered because the shape is constantly changing and the top cannot be kept convex.

A simple Dutch barn on a steel, timber, or concrete frame is relatively cheap to erect. If it is sited in a sheltered spot there may be no need for side cladding; on a windy site the rain will blow in horizontally from the eaves, particularly when the barn is nearly empty. Such a barn is easy to load and unload and may be adapted for other uses when empty. The old-fashioned three-bay barn – the centre bay with doors each end for loading – has much in its favour. It offers more protection from wind and the centre bay can be used as a cart or tractor shed. In Wales one empty bay is normally used as a holding and catching pen and the centre bay becomes the shearing shed. On my smallholding, I have lofts above the cow-byre and have built a traditional-style three-bay barn of concrete blocks and slate roof. The concrete walls are six feet high and from there to the eaves spaced timber cladding breaks the wind and gives good ventilation.

Granary

Threshed grain and other concentrate feeds must not only be kept dry, but also be protected from birds, mice, and rats. They may be stored in bins or sacks, loose on the floor, or in silos. On my farm I use three methods; most of my bulk oats are stored in a ferrocement silo which is completely weather- and vermin-proof.

Any surplus is stored in bags on the wooden loft floor above the cow-byre. A cat and the occasional use of warfarin keep the rats and mice more or less in check. Rolled oats, bran, and other ready-to-feed cereals are stored in 44-gallon drums with lids. It is most convenient to have a relatively small room near the byre for mixing and storing feeds in the short term.

Harvesting roots

The harvesting of roots is not as critical as that of hay and cereals: apart from the risk of frost, the crop will not be damaged by bad weather. Nevertheless, there can be serious problems carting in a wet autumn. When stored, roots and potatoes need protection from frost, waterlogging, and farm animals. So long as they are on a well-drained surface above ground-water level, moisture from above will not harm them, and clamps may be built outside anywhere sheltered from strong winds. For ease of access and feeding, it may be more convenient to store in a building or under a lean-to roof.

Mangolds are very sensitive to frost, and must be lifted, carted, and clamped by November. Like all the beet family, they bleed easily, and should therefore be topped leaving some of the stalk on the bulb and the roots should be cleaned but not trimmed. I use an old butcher's knife; the sharp edge for topping and the back of the blade for cleaning the roots. The mangolds are piled in small 'tumps', covered by their leaves and left for a week or ten days to sweat before they are carted. These tumps should be placed in rows with plenty of room for a tractor and trailer to move between them. They are later carted to a 'clamp' which should be sited on free-draining ground, and covered with up to a foot of straw or bracken. This, in turn, may be covered with a layer of earth or a tarpaulin, and the clamp surrounded by a shallow ditch to prevent waterlogging. The mangolds will remain alive in the clamp, and the stalks will be sprouting by March. They mature in the clamp, and are normally fed from February to June.

Common turnips grow very quickly, but are neither frost-hardy nor good keepers. They should either be grazed *in situ* in the autumn or early winter or, if clamped, fed before Christmas. A tougher variety known as 'hardy green turnips' are generally sown with rape for autumn and winter grazing.

Swedes, which are much more frost-resistant, can be left in the field until Christmas and grazed *in situ*. They will continue to grow wherever the ground is slightly above freezing. They may be lifted and topped whenever convenient, using a sharp topping knife to trim stalks and roots quite drastically to get rid of surplus soil. They may be carted as soon as they are lifted. In the clamp they last well until February, but tend to rot when the weather warms up in the spring. They are normally chopped and fed from December to mid-February. I usually lift and clamp the best swedes and leave the smaller and damaged roots to be grazed *in situ* by sheep.

6 Livestock

The choice of livestock enterprises on a smallholding is governed by a number of factors, First, the size of the holding: the smaller the farm, the narrower the range of animals it can support from its own resources. While a 50-acre farm may manage a viable dairy herd or a wide range of animal projects, on ten acres a house cow and a few sheep or pigs would be the limit. On two acres it would be unusual to find a cow at all, and on half an acre a smallholder would be limited to a few chickens, a goat, a pig, or rabbits. It is possible, of course, to run intensive livestock enterprises (poultry, pigs, veal, and so on) on very small holdings, buying in all the food, but that is factory farming, and not within the scope of this book.

The second factor is the location and nature of the holding: dairy cattle will not pay in an area subject to long periods of summer drought, nor can store cattle be fattened on marginal hill land.

The third limitation is the proximity of a suitable market for animal products and the profitability of various classes of livestock, which fluctuate both seasonally and over the years. Finally your own 'delight' in any class of livestock will play an important part in the system you adopt.

Whatever animals are chosen, it is important that the smallholder should know the long-term stock-carrying capacity of his farm and resist the temptation to overstock. Bought-in feeding-stuffs soon make off with profits, and over-trodden fields quickly reduce fertility and production.

To maintain good swards and healthy animals, monostocking (with one species only) should be avoided. Different animals graze in different ways with different effects on herbage: mixed stocking therefore produces the best balance of grasses and clovers. Each species suffers from its own series of parasitic worms and is unaffected by others; so mixed animals grazing together greatly reduce the overall parasitic burden.

Whatever the size of the farm, it is highly desirable to have a source of animal waste to maintain the humus level in the soil: to sustain fertility on a stockless holding is very difficult. It is sometimes possible to import organic manures from neighbouring farms or stables, particularly on a small scale. The larger the holding, the more desirable it is to keep livestock as an integral part of the balanced whole.

Shelter

Apart from the intensive management of poultry, pigs, and veal calves, which finds no place in this book, there are four main reasons for providing shelter for farm animals: first, to reduce stress in foul weather; second, to maintain condition and production with the minimum feed; third, to protect the soil structure in the fields from physical damage; and fourth, for the convenience of the farmer when handling, feeding, milking, or lambing. Over-protection or cosseting of farm animals, however, destroys their hardiness and should always be avoided. Running truly free-range on well-drained, low-stocked land, most animals will find their own shelter from the wind and retain health and hardiness. In bitter weather they need more food, their condition and production dropping in proportion to their stress. For eight months of the year cattle and sheep will certainly be out of doors, coming in only for milking or handling. Pigs and goats may be part or full time in or out according to the system of management. Poultry will certainly be shut up at night as protection against foxes and may be more or less restricted by day. Most sheep live out the year round, although an increasing number of flocks are now coming into straw yards or slatted floors for the worst of the winter and for lambing. During the four worst months (mid-December to mid-April) grass is scarce and what there is has little value. The ground is wet and liable to poach, and the farmer has good reason to bring his cattle in.

In-wintering of livestock carries the risk of ill-health: respiratory ailments due to bad ventilation, diseases of the gut (particularly scour in calves) due to bacteria which are difficult to clear from infected buildings, and diseases of the feet and limbs due to standing in muck and lack of exercise. New-born calves and lambs readily pick up bacteria via their navels, and are particularly vulnerable to indoor risks. It follows that buildings used for housing stock should be very well ventilated without being draughty, easy to clean and disinfect, well drained, and with ample room for exercise (unless a separate exercise yard is provided). A clean water supply and suitable feeding mangers or racks must be provided. Access for mucking out is also important.

Stock-handling

The quiet and efficient handling of farm animals is the mark of a good stockman, and cannot be accomplished without proper facilities. Catching, restraining, injecting, dosing, de-horning, foot treatment, nose-ringing, castrating, dipping, and shearing are all simple operations in the right conditions, but can be difficult or downright dangerous to man and beast without them.

The first requirement is a strong stock yard with one or more pens into which animals may be sorted. A simple cattle crush may be built into the yard fence in such a way that beasts naturally walk through it in their daily routine and are unafraid when it restrains them.

Next to the stock yard are the sheep-handling pens, dipper, draining pens, and shedding race. If cleverly sited in relation to existing fences and walls, the sheep will run freely through the race and can be sorted without handling.

The efficient shearing of a flock of sheep depends upon the layout of the shearing shed. As this operation takes place on only one or two days a year, no smallholder can justify a permanent shed; he must adapt some part of a multi-purpose building. The basic covered requirements are holding pens, catching pen, shearing floor (or benches) with power points (unless you're using hand shears), fleece-rolling table, and space for the wool sheets. Sheep can be shorn out of doors, but rain can spoil a carefully planned shearing day.

Cattle

There are several enterprises suitable for smallholdings, depending upon acreage, soil, climate, and the farmer's bent. They include the following:

- a small commercial dairy herd for the sale of whole milk or of cream, butter, cheese, or yoghurt
- a house cow to meet domestic needs
- calf-rearing of dairy heifers or beef on nurse cows or by artificial means
- storing dairy replacements or beef animals
- fattening beef steers or heifers.

Cattle breeds

The ideal dairy cow is one that will give a good yield of high butterfat milk over a long lactation at low cost. She will be of hardy constitution and produce quick-maturing calves (either pure or cross-bred) that are readily saleable in the local market. She will have a quiet temperament, although I believe that this is usually more a matter of environment and handling than of breeding. The ideal nurse cow will be of beefier type, milky, but with less emphasis on butterfat.

The choice of breed will depend largely on local conditions, market preferences, the size of the smallholding, and, most of all, the aims of the farmer.

The Friesian is the highest-yielding dairy breed. It is a large animal requiring heavy feeding. The milk is of low-butterfat content. Pure-bred Friesian calves make good beef; crossed with a beef bull, excellent beef calves are produced. On good land, Friesians might be considered by a smallholder for a milking herd, but would not be suitable for a butter and cheese project or as smallholders' house cows.

The Shorthorn is also a large, high-yielding cow with largely the same characteristics as the Friesian in so far as the smallholder is concerned. The Shorthorn is more variable in conformation, some being primarily dairy and others pure beef types.

The Ayrshire is a smaller cow with lower food requirements and higher butterfat content than the Friesian or Shorthorn. It is a hardy breed that does well on poor, cold land. Under favourable conditions it can be a high yielder, and the milk is very favoured for cheese-making. Pure Ayrshire calves do not make good beef, and as such are not so profitable to rear or sell. Crossed with beef bulls they sometimes produce adequate beef calves.

The Guernsey is a fairly hardy cow of medium yield with a high butterfat content. They are very suitable for a small-holder's dairy herd, for a cheese and butter enterprise, or as house cows. Like Ayrshires, their calves seldom produce good beef.

The Jersey is a smaller, more compact cow than the Guernsey with a very high butterfat milk. Her yield is variable, but can be remarkably high for such a small cow. Pure Jersey calves are considered virtually valueless for veal or beef, although I have reared such calves to 14 weeks on nurse cows and produced excellent veal carcases. Cross-Jersey calves are variable in conformation, and generally disappointing, although I have seen good beef calves out of old Jersey cows by Limousin or Charollais bulls (I would not risk so large a bull on a young Jersey). The Jersey cow has possibly the most docile temperament of any breed (Jersey bulls are another matter) and is my favourite house cow. Her food requirements are very modest, and properly acclimatized, she is extremely hardy. A great advantage is her willingness to be tethered to graze in garden or orchard.

The Kerry is a very small beast of dairy type and extremely hardy constitution. Its yield is satisfactory in proportion to its size and food requirements, and the butterfat content is fairly high. It is a useful house cow for a very small holding, and, crossed with an Angus, produces good beef.

The Dexter, like the Kerry, is a very small beast of Irish origin. It is more stocky than the Kerry with very short legs and large head. Like the Kerry, it has a useful yield of high-quality milk and a small appetite. It is very hardy. It produces good, small beef-type calves. Monstrous calves of the so-called 'bulldog' type are much commoner than in other breeds. Indeed, the Dexter appears to be a heterozygous form, producing, when bred pure, 25 per cent of long-legged 'Kerry-type' calves, 50 per cent of true Dexters, and 25 per cent of 'bulldog' calves.[9] Like the Kerry, it is a good house cow for the smallholder.

The Welsh Black is a good dual-purpose beast, producing excellent beef calves and a good yield of medium-quality milk. The fat globules are very small, and more difficult to make into butter than the cream of Channel Island cattle. Good Welsh Blacks have too much milk for their own calves, and make good nurse cows. I have found that a Welsh Black is a useful addition to a Jersey or Guernsey house cow, whose calf she takes over, in addition to her own, from about one week old. If the calving dates are right, the Black may give some milk for the house when the Jersey is dry.

Cross-bred Cattle: dairy–beef crosses, notably Hereford x Friesian, Hereford x Welsh Black, Angus x Shorthorn, and Shorthorn x Galloway (Blue-Greys), make excellent dual-purpose cattle, and can have a place on the smallholding as house-cum-nurse cows.

Exotic breeds such as the Charollais and Siemental have recently found much favour for crossing with dairy cows to produce top-quality, quick-maturing beef calves. Care must be taken, however, to avoid putting such large bulls on young, small cows, lest serious calving troubles result. Not only may small cows have difficulty delivering the calves, but may overstrain themselves in producing and feeding such demanding offspring.

Recognition of heat

Sexual maturity in cattle is reached at about one year of age; earlier in Jerseys and later in hill breeds, but generally some months before they are fully grown. If she is put in calf too soon, a heifer's growth and productivity will probably be stunted. A well-nourished cow will come in season regularly at intervals of 19–23 days, the 'heat' lasting 12–24 hours in summer but for a shorter period in winter. The best time for service is towards the end of the heat.

A cow in season is restless, looks about her with a more alert interest, tries to mount other cattle *and will stand still when mounted*. She may bellow incessantly, generally shows a slight swelling of the lips of the vulva, and produces a clear slimy discharge. She may also bear marks (grass, mud, and so on) over the tail-head and flanks where she has been mounted. A sudden drop in milk yield for twenty-four hours and scouring in the calf are also symptoms of heat. A smallholder with only one cow may have difficulty in spotting the heat. He should keep notes in his diary and watch carefully from the 18th to the 24th days for any of the symptoms. If in doubt, it may be wise to walk his cow to a neighbour's farm to meet other cattle.

A cow that has been allowed to get into very poor condition may thereby have suffered damage to her ovaries, and will be less likely to come in season. An overfat cow is also likely to be less fertile, and an old cow may cease to breed altogether.

A smallholder will seldom find it worth keeping a bull. You can either use artificial insemination or take your cows to a neighbour's farm. When using AI it is particularly important to keep a close watch on your cows to see exactly when they start bulling: service is most likely to be successful towards the end of the heat.

Calving

The average gestation period for a cow is 284 days, but it is not uncommon for calving to occur up to ten days early or late. Channel Island cows are more likely to be late than other breeds.

From about the sixth month of pregnancy it is often possible to see the calf moving. As calving approaches the udder begins to swell and fill, the vulva becomes grossly enlarged and the muscle and cartilage connecting the pin-bones to the tail-head soften, allowing a sunken area to appear on either side of the tail-root.

As calving becomes imminent, the cow's back becomes arched, the tail is raised, the ears are laid back, and the eyes bulge. She may seek a secluded spot away from other cattle. As labour begins, the water bag will appear; this will soon burst, the cow licking up the contents as a stimulant to her mothering instincts. In a normal birth the forefeet will soon appear, followed by the nose. At this point the stockman can assist by pulling downwards (towards the cow's hocks) on the forelegs in rhythm with the cow's labour, using ropes if necessary. As the head appears, mucus should be cleared from the nostrils and mouth. The top of the head and the shoulders are the most difficult parts to deliver; once they are clear of the pelvis, the rest follows easily.

When the calf is delivered, it should be left for half a minute with the navel cord still connected to ensure the maximum blood flow from the cow. The cow should get up and lick the calf to dry it and stimulate circulation. The calf should get to its feet and search for milk within a few minutes of birth.

The pros and cons of indoor versus outdoor calving will long be debated. Indoor calving is more convenient for the farmer because it reduces damage to fields in bad weather, and eliminates the risk of the calf falling into a ditch or stream. Nevertheless, it is now known that an outdoor calf makes better use of colostrum than one born in a well-littered loose box, and better still than one born in a byre. The cow's freedom to lick and nurse her calf in natural surroundings seem to have a marked effect on its well-being. The danger of bacterial infection leading to white scour becoming endemic in buildings is very well known.[10]

The afterbirth is normally expelled within 12 to 24 hours of calving. If it is not cleared within 72 hours the vet should be called. It should not be pulled by an unskilled cowman, as it may break off leaving a portion attached to the uterus which will cause infection.

Malpresented calves (legs back, head back, breech-first, for example) and oversized calves may have to be manipulated and pulled. The smallholder should always have a pair of calving ropes available, but should call in a vet until he has gained wide experience of calving problems. Many farmers believe that regular exercise before calving reduces the risk of malpresentation – a further reason for outdoor management.

Drying off

The lactation period in cattle varies greatly. Some beef breeds (notably Herefords) have short lactations and dry themselves off in about 200 days. The dairy breeds, on the other hand, must be deliberately dried off to ensure a rest period before calving. The normal dairy lactation is 305 days, but in some cases, a cow will continue to milk right up to calving, or, if not in calf, for fifteen months or more unless she is forced to dry off. If the yield is very small, milking may be stopped abruptly, the cow put onto meagre, dry rations for a few days, and the milk will be reabsorbed and cease to flow. If the pressure builds up enough to cause pain, it will be necessary to ease it once or twice.

In a higher-yielding cow, I prefer to dry off over a period of two or three weeks, milking once a day for about ten days, and then easing only if the pressure builds up.

Cattle housing

For cattle there is a wide choice of systems. The byre, where beasts are tied by the neck; loose housing on deep litter in shed or covered or open-topped yard; and cubicles (individual stalls where cattle are free to go in or out) are all used for adult and yearling cattle. Smaller loose boxes are favoured for calves. To be hygienic a byre needs a raised standing and a dung-passage, the distance between tie and passage varying with the breed (4 ft. 6 in. for Jerseys, 5 ft. 6 in. for Friesians). There must also be partitions to contain pairs of cattle to prevent them turning tail-to-manger. Byre-tied cattle must be turned out for exercise and water. Loose housing is the simplest system, and most satisfactory for 'dry' cattle; but milking cows tend to get very dirty. Cubicles, littered with sawdust, shavings, tan bark, or paper are dry, warm, and economical. Their design and size is critical in relation to the size of the animals. They are widely used with milking bales, where the cattle come into a few stalls in succession to be fed and machine milked, and with self-feed silage systems in large dairy herds. Loose boxes for calf-rearing can be easily made in existing buildings, using hurdles, wooden partitions, or straw bales.

A smallholder with a commercial dairy herd could use cubicles with milking bale or the byre. For a smaller unit (say one house cow and one or two nurse cows) he would probably prefer the byre. On my own farm in winter my two cows are tied at night and milked in the byre, and are turned out in an exercise yard by day. The calves are on deep straw litter in an adjoining loose box, but turned out to graze in the middle of the day when the weather is kind and the ground reasonably dry. The most important features of the exercise yard are shelter from the wind, hard standings under the hay racks to avoid the area turning into a quagmire, a good water supply, and enough space for exercise.

Dairy cattle

A dairy herd for selling milk is a very common enterprise on conventional smallholdings. To milk twenty cows with modern equipment is well within the capacity of a farmer and his wife, although there can be few weekends off, much less longer holidays. The great advantage is the regular milk cheque which, provided the profit margin is adequate, gives security within which you can plan. If you are a good stockman you can, as a smallholder, have a great advantage over the large farmer; through intimate knowledge and personal touch, you can get high production at low cost without risk to the health of your herd.

A milking herd for the sale of cream, butter, yoghurt, and cheese is a more specialized enterprise requiring additional skills and facilities and a retail market. It also entails complementary livestock (poultry or pigs) to use up the skim or whey. In skilled hands, this can be more profitable than selling to the Milk Marketing Board, but involves even more labour.

Many smallholders will keep only one or two cows to be hand milked for the house and to rear calves for sale. I have done this for many years, keeping a Jersey for milk, butter, yoghurt, and cheese and a Jersey x Hereford nurse cow to rear her own and the Jersey's calf. The whey is either boiled down to make Norwegian-style whey cheese or is fed to poultry or pigs. Sometimes, when the Jersey is dry, it is possible to take a little milk from the Hereford for the house.

Feeding for milk production

A dairy cow's ability to convert food into milk depends upon her inherited milking qualities, on the way she is reared from weaning to maturity, and upon her feeding when in calf and in milk. It should go without saying that a good dairy cow puts her food 'into the bag' rather than 'on her back', and it is common to see the best milkers in very lean condition. It is well known that slow-maturing dairy heifers, reared on nurse cows and a relatively low ration scale, milk better and live longer than those reared quickly like beef calves; indeed a dairy cow carrying too much 'condition' is unlikely to be amongst the best milkers.

Cattle need food for two purposes: first, to maintain their bodies and normal activity (the maintenance ration); and, second, to produce milk or meat or to develop the unborn calf (the production ration). Up to the time she is served, a heifer, in addition to her maintenance requirement, needs a protein-rich production ration to grow to her mature conformation. From first bulling to calving she needs an additional ration for the growth of the foetus (she may still be growing herself) and to build up her reserves to meet the heavy demand of early lactation. From calving (when she will be at her maximum weight) to the peak of her lactation some six weeks later a cow will be milking 'off her back', using the body reserves built up while she was in calf. A cow inadequately fed during pregnancy will still build up to a peak of lactation to meet the growing needs of her calf, but will sink badly in condition and dry up earlier as a result. From the peak of lactation, the milk yield will drop gradually (about 12 per cent per month) while the cow's body weight will recover to reach a maximum at the subsequent calving. Feeding, therefore, should be planned to cope with the whole lactation and pregnancy; it is of little use to try to rectify a low yield, poor body condition, or early drying-off by extra feeding if the buildup of reserves at the right time has been neglected.

Feeding-stuffs are classified as concentrates, succulents, and roughage. Concentrates–including cereals, linseed, and fish-meal–are high in energy and/or protein, low in moisture and roughage, and are highly digestible. Succulents–including grass, silage, roots, and green crops–are high in water content and energy but low in starch equivalent: the dry matter is very digestible. Roughage–including hay and straw–is high in fibre content but relatively low in energy and protein. It is less digestible than concentrates and succulents.

Midsummer grass contains sufficient fibre to meet a cow's requirements, but during the spring and autumn flush and in winter, extra fibre (hay or straw) is required. Lack of fibre results in a low butterfat content in the milk.

In winter it is necessary to provide succulents (roots, kale, or silage) to keep the intestine content reasonably soft and provide energy without overfeeding on concentrates.

Concentrates may be fed to provide energy and protein, especially in the last weeks of pregnancy to build body reserves and to feed the growing calf at a time when there is not enough room in the rumen to take the full requirements in succulents and roughage.

Full details of feed values and requirements are shown in

Table 4. Hay equivalents

1 kg. average hay	=	3 kg. silage
,,	=	4 kg. kale, beet tops, or swedes
,,	=	5 kg. mangolds
,,	=	2 kg. barley straw
,,	=	0.5 kg. dairy cake (concentrate)

Russell's *The Herdsman's Book*; sufficient here to quote two simplified tables from it as a general guide (Tables 4 and 5).

Table 5 must obviously be adjusted by the stockman to meet the varying requirements of pregnancy, lactation, and buildup of body reserves during the different phases of the breeding and production year. For my own April-calving Jersey house cow, yielding up to 18 litres a day (3,636–3,864 litres over a 305-day lactation), I feed roughly the following:

Table 5. A cow's requirements in terms of hay equivalent

Breed	Liveweight (kg.)	Requirement	
		Daily Maintenance	Production of 1 kg. milk
Friesian	560	9 kg. hay equivalent ⎱	
Ayrshire	460	7 kg. hay equivalent ⎰	0.35 kg. dairy cake
Jersey	380	5.4 kg. hay equivalent	0.5 kg. dairy cake

March to May (calving in April)

    A little grazing when available

    10 kg. chopped mangolds

    4.5 kg. best hay

    1 kg. rolled oats

    0.25 kg. seaweed meal for minerals and trace elements

June to September (in full lactation, in calf)

    Ad lib grazing for maintenance and 4.5 litres milk

    Up to 2 kg. hay to make up roughage lacking in early grass (sometimes until mid-June)

    Up to 1.5 kg. rolled oats for up to 14 litres milk

    0.25 kg. seaweed meal

October (falling lactation, in calf)

    Up to 3 kg. hay (or 4 kg. dredge corn on the straw) to supplement sinking vegetation

    1 kg. rolled oats for waning production (not if dredge corn is being fed)

    0.25 kg. seaweed meal

November–December (falling lactation, heavy in calf)
  5 kg. undersown dredge corn on the straw (oats, barley, and peas)
  6 kg. chopped swedes
  0.25 kg. seaweed meal
January–February (dry, very heavy in calf)
  3 kg. straw chaff ⎫ mixed
  8 kg. chopped mangolds ⎭
  4 kg. hay
  1 kg. rolled oats
  0.25 kg. seaweed meal

Some authorities, particularly those intent on achieving exceptionally high yields, advocate a 'steaming-up' programme during the last six weeks of pregnancy in which an increasing quantity of balanced concentrates is fed, based on the estimated peak yield; reaching a maximum of about 1 kg. of concentrate per 5 litres of milk. This certainly produces the milk, but can cause inflamed and pendulous udders before calving.

I much prefer to aim at an 'optimum' production in which the cow is not forced and complications are less likely. To achieve this I try to bring the cow more gradually to top condition at calving by basing my feeding on the whole breeding and production year (see Figure 14).

Lactation and milking

Occasionally a high-yielding cow will develop a hot, hard, and distended udder two or three days before calving, particularly if she has been 'steamed up'. It may be necessary to ease the pressure by milking and soften the teats and udder by massage. A cold water hose for five minutes twice a day will lower the bag temperature and reduce the risk of a sudden multiplication of mastitis bacteria. The milk withdrawn before calving is colostrum, which is needed to protect the calf. It should therefore be saved and put in the refrigerator. Colostrum is produced for three or four days after calving,

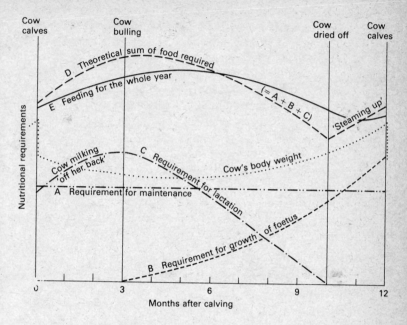

Fig. 14. A cow's nutritional requirements through the breeding and production year

gradually being replaced by milk. The udder and teats are often under great pressure, and one is tempted to strip the cow out after the calf has sucked. This is normally unnecessary, and may even bring on milk fever (caused by a sudden reduction of the available calcium in the bloodstream which occurs in heavy milking immediately after calving). If mastitis is suspected, however, the udder may be stripped as soon as the calf has sucked its fill, the colostrum being given back to the cow to drink. By this means mastitis can be spotted and treated before it becomes serious. Once the milk is free of colostrum, the cow should be milked out normally.

Milking should be carried out at regular times; ideally at

twelve-hour intervals. Cows are creatures of habit, and will let down their milk more readily on a completely regular routine. The udder should be washed thoroughly in warm water and massaged at the same time to trigger the hormones which let down the milk. The cow should then be fed and you should start milking within a minute or two of washing. In herds of several cows it is wise to include a mild disinfectant in the water; with a single house cow, being hand milked, this is not so vital so long as utensils and the milker's hands are scrupulously clean. I have found that hyperchlorite (a most common disinfectant) tends to cause teat chapping.

One jet of 'foremilk' should be taken from each quarter, into an inspection cup (or onto a black wellington boot) and checked for any sign of mastitis lumps.

The cow should be milked out and stripped as quickly as possible before the 'let down' ceases; if this is not done, she may develop lazy habits and become difficult to milk out. When milking by machine it is extremely important not to leave the cluster on when the milk flow has ceased, or inflamed teats and mastitis will quickly follow. A cow with an uneven udder or a damaged quarter is particularly vulnerable to machine injury, although perfectly all right when hand milked. Indeed, I have bought excellent three-quartered house cows very cheaply and found their yield practically unimpaired. When hand milking it is very easy to notice the slightest change in the udder or teats and to take the necessary action. When stripping, each quarter should be massaged between thumb and fingers while the teat is emptied by a slight downward pressure on the teat cup (or by pulling between thumb and forefinger when hand milking).

Restraining difficult cows

If calves are handled from birth, there should be no subsequent difficulty in getting them to stand still for milking unless they are in pain. Sometimes teats and udder become swollen and tender after calving, in which case it may be necessary to restrain a cow whilst she is eased or milked. Restraint may also be necessary when fostering a strange calf onto a nurse cow.

The simplest method, generally effective, is to pass the tail forward between the hind legs and hold the end firmly above the hock. If the cow tries to kick, she does so against her own tail, and soon gives up. A more positive method is to tie a rope quite tightly over the back and under the belly, just in front of the udder. If the cow kicks, the rope tightens and hurts her. Leg hobbles may be used, but these are liable to make a cow panic and hold up her milk. If all else fails the cow should be sold.

Calf rearing

The production of mature dairy or beef cattle takes place in four stages, one or more of which may be undertaken by the smallholder, depending upon his land, the climate, and the housing available. With the right facilities, calf-rearing may be a major farm enterprise, in other cases the smallholder will merely rear an occasional house cow replacement or attempt to make the most of his home-bred calves as an inevitable by-product of dairy or house cow.

Stage one is the breeding of suitable calves, for veal, beef, or dairy heifers. (The breeding and rearing of bulls is a very specialized skill requiring ruthless selection and culling, and is thus inappropriate to smallholding.) These calves may come from pure beef, dairy, or cross-breeds, depending upon the smallholder's arrangement of enterprises. Stage two is the rearing of calves from birth to weaning. Stage three is the 'storing' of calves from weaning through the growing period; and stage four is the finishing (or fattening) of beef, or the preparation of a heifer from bulling to calving.

The breeding of calves specifically as dairy replacements would be undertaken only as a by-product of a dairy herd. As

50 per cent of the calves would be males destined for veal or beef, a smallholder is unlikely to use a dairy bull except on beefy-type cows such as Friesians, Shorthorns, or Welsh Blacks. He would be more inclined to cross his cows with a beef bull to produce the most saleable calves.

The rearing of beef or dairy calves from birth to weaning can be a smallholding enterprise on its own, selected calves being bought in to be multiple suckled on nurse cows or bucket fed. Wherever the calves come from, it is essential to ensure that they have had ample colostrum during the first twenty-four hours of their lives (since this is impossible to tell at market, you would be wise to buy calves only from farmers you know). After that time they are unable to absorb the colostrum proteins. Without colostrum, calves have little or no protection from diseases.

If you are using the *multiple-suckling* method a good, milky nurse cow may be expected to rear from three to eight calves in one lactation. There are two main systems: in the first, the same three or four calves remain on the cow throughout the lactation; in the second, three successive batches of calves are suckled for about three months. In all cases it is most important to ensure that the calves receive colostrum at birth. Hay, concentrates, and/or grass are offered to the calves from one week old to develop the rumen. Beef calves require a higher plane of nutrition than dairy heifers.

Multiple suckling, as a smallholder's enterprise, relies on bought-in calves, which are most readily available and cheapest in the autumn. A common system is to buy in an autumn-calved cow and calf in September – a dairy-herd cull with a lost quarter or a slow milker would be quite suitable – rear one batch of calves before Christmas, a second by March, and a third during the early part of the grazing season after which the cow can be expected to fatten for slaughter or be kept on if safely in calf. Multiple suckling requires some accommodation for the calves; loose boxes or straw-bale pens are adequate. The third batch may be reared in the field in early summer. It is most important to feed the nurse cow to dairy standards to ensure high production and a long lactation.

On my own smallholding my two cows (a Jersey house cow and a very beefy Hereford x Jersey) are both spring calvers: I foster the Jersey's calf onto the nurse cow and occasionally buy in a third calf. For the first six to eight weeks, the calves are kept in a loose box and turned to the nurse cow twice a day to suck. Meadow hay, a little rolled oats, and seaweed are offered from the fifth day. At about eight weeks (or when the grass is growing and the calves are sucking freely), they are turned out with the nurse cow for the summer. My nurse cow produces a lot of milk in the first ten weeks, but then rapidly dries up – a typical fault of some strains of Hereford. As a result she has to be watched carefully lest the calves are unable to cope with the milk in the early weeks.

Bucket feeding requires specialized accommodation in the form of pens or individual stalls as well as food-mixing and sterilization facilities. It is also more labour-intensive than suckling. Strict hygiene and a regular routine are essential. An artificial udder, placed so the calf's head is held high while sucking is infinitely better than a bucket at floor level. With the latter, some of the milk gets into the rumen (the first stomach) instead of into the abomasum (the fourth and true digestive stomach) where it can be properly clotted and digested. This is because the entrance to the rumen is open when the calf's head is lowered, and closed when the head is raised.

Whole milk is gradually replaced by milk substitute, concentrates, and hay from the fourth to the eighth weeks, and the calves are weaned onto dry food alone by the twelfth week. A calf reared on whole milk – especially if suckled – is likely to be healthier and longer-lived than one reared on milk substitute.

In the *early-weaning* system whole milk is replaced by milk substitute after ten days and the calves are weaned onto dry food in seven weeks. The calves are introduced to hay and a special high-protein concentrate from the fifth day.

Bought-in calves are liable to have gone through a period of stress and starvation as well as exposure to disease in the market. It is therefore vitally important that they should not be allowed to gorge themselves for the first few feeds on arrival at the smallholding. Good hygiene in buildings and feeding equipment will reduce the risk of the deadly bacterial (white) scour.

Beef production

Calves destined for beef, in addition to being of the right 'type', should show evidence at weaning (that is to say at six to eight months) of good nutrition throughout calfhood. Small, pot-bellied, thin-backed calves will not recover to make high quality, profitable beef.

Well-fed calves of the right breeding and conformation may either be fed straight on to make 'baby beef' at fifteen to eighteen months or may be stored through the winter in relatively hard conditions and on low nutrition, to be fattened cheaply on grass through the following summer and autumn. The quick-maturing Angus, Hereford, Beef Shorthorn, and some exotic crosses feed well as baby beef. The slower-maturing, hardy mountain breeds are better stored; they then make very rapid weight gains on grass.

Cattle may be stored indoors (tied or in loose sheds), in sheltered yards, or out of doors in well-drained fields. The latter system keeps them in good health and they go on to fatten without a check when the grass starts to grow in the spring. A 254 kg. store may be fed on $3\frac{1}{2}$ kg. of barley or oat straw, 18 kg. of swedes or mangolds, and a small supplement ($1\frac{1}{2}$ kg.) of rolled oats, this being sufficient to give rather more than $\frac{1}{2}$ kg. a day weight gain.

A smallholder with adequate facilities (sheds, yard, or dry fields) could do well to buy in a few six- to eight-month-old bullocks in the autumn, store them through the winter, fatten them on grass through the summer, and then sell them (either fat or for finishing) before Christmas.

Well-grown stores, from twelve to eighteen months old, may be fattened on grass, or indoors on the by-products of arable farming. Fattening pastures, as described in Chapter 2, require rather special conditions of soil and climate, whilst profitable indoor fattening depends upon the availability of cheap feeding-stuffs, most likely to be found as by-products on a larger, arable farm. Only if these conditions are met, would a smallholder be advised to buy in beasts for fattening.

Sheep

If well managed, sheep can be a profitable enterprise on a smallholding, either as a source of meat and wool for the family, or for the sale of lambs, meat, wool, or breeding stock.

Modern British sheep are reared primarily for the production of fat lambs: the sale of rams, breeding ewes, and wool is generally a by-product of the main lamb-producing enterprise and is of lesser importance. However, an expert producer of top-quality rams may find them to be his most profitable venture.

It is very important that you should select a breed that suits your land and climate, while at the same time providing you with a good profit. Ideally one would like a hardy, prolific, milky type of ewe with a strong mothering instinct, producing quick-maturing lambs of ideal butcher's conformation and a heavy fleece of high-quality wool. Unfortunately these qualities seldom go together, the farmer being obliged to find the compromise best suited to his particular conditions.

The hardiest sheep are the small mountain breeds that thrive on very sparse grazing in extreme weather conditions.

They breed late in the season, have a low lambing percentage – producing singles rather than twins or triplets – and are slow maturing. Their lambs are very small but generally of high-quality meat. When transported to strong land they tend to breed earlier and are more prolific, but can become overfat and lose their hardiness.

At the other extreme, the heavier breeds require considerably better grazing, breed earlier with a long breeding season, are more prolific, quick maturing, and less hardy.

Between these extremes is a wide range of pure breeds and crosses developed over the years to meet special markets to suit particular conditions. These are described in detail in the booklet *British Sheep*, published by the British Sheep Breeders' Association.

Lambs

On good land, most smallholders will aim at selling lambs fat off the ewes in the autumn without additional feeding. In particularly good conditions it may be worth the extra trouble involved in very early lambing – in January or even late December – to have lambs to sell at a premium by midsummer.

On marginal land and under harsh conditions, the later-born lambs may have to be fed from weaning until Christmas or sold in the autumn as stores. In recent years the demand for lean meat and small joints with little or no fat has completely altered the specification of the ideal 'butcher's carcase': lambs that would earlier have sold as stores are now graded as fat.

The most profitable lamb is one that grows and fleshes up quickly and steadily on an abundance of milk and clean, nutritious grass. Any check in this growth caused by poor nutrition, accident, or disease results in a less valuable lamb, in higher costs, or both. It is therefore important that newly-lambed ewes should have access to fresh grass to stimulate and maintain their milk flow, that the grazing be rotated onto clean pastures to avoid the buildup of stomach worms, and

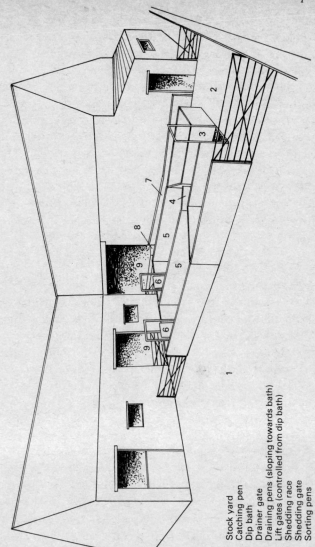

1 Stock yard
2 Catching pen
3 Dip bath
4 Drainer gate
5 Draining pens (sloping towards bath)
6 Lift gates (controlled from dip bath)
7 Shedding race
8 Shedding gate
9 Sorting pens

Fig. 15. Outline of sheep-handling facilities tailored to fit the shape of existing buildings and yard

that foot-rot and fly-strike be treated promptly before they affect the lamb's vigour and appetite. Handling and dipping facilities are illustrated in Figure 15.

Late-born lambs may have to be weaned before they are ready to sell fat or as stores; to leave them too long on their dams at the end of the grass-growing season is to risk the ewes going to the ram in sinking condition which may affect not only their fertility, but their ability to pick up condition and rear good lambs the following year. Provision must therefore be made for supplementary feeding. The aftermath of hay fields, laid up until September or October, rape and turnips sown in July or August, or hand feeding with concentrates are alternatives open to the flockmaster, who must weigh many factors and judge which course will be most economic.

Breeding stock

Provided you buy in well-bred foundation stock, you should be in a good position to produce a few top-quality rams and breeding ewes for sale. It goes almost without saying that inherited factors put a top limit on the potential quality of any animal; nevertheless, the development of its *full* potential is a matter of sound stockmanship and attention to detail. Here the small farmer should be at a great advantage.

The sale of rams and potential show-class yearlings and lambs, expertly reared and produced, can be very profitable for the farmer who has the time and the 'delight'. The selection of ram lambs to keep for breeding requires a good eye and experience of the breed. A large, early-born single may easily be chosen in preference to a smaller twin of higher potential. Knowing your sheep as individuals, you should be able to look beyond the ram lamb to his dam and her progeny, to her twinning record and her health as well as to the sire, before deciding whether to keep or castrate him.

Hill ewes, two to five years old, sound in tooth and udder,

are much in demand for cross-breeding on stronger land, where they can remain profitable for several years more. The hill farmer can draft from his flock good ewes that do not, however, meet his particular requirements or are becoming too old to thrive in hill conditions.

Wool

The breeding of most British sheep has concentrated for many years on the economic production of fat lambs. Wool, though an important by-product, is no longer of such prime concern. Nevertheless, some of the old pure breeds have very fine fleeces which find a ready market amongst the growing fraternity of hand spinners. Notable amongst these are the Shetland and Jacob breeds, producing the finest, coloured wool, although the quality declines when these breeds are brought onto strong land. If you keep up to half a dozen high-quality ewes you can sell your wool at a considerable premium, but owners of larger flocks are obliged by law to sell to the Wool Marketing Board at the standard price for the grade concerned.

Whether selling privately or to the Board, it is important to present the fleeces to the best advantage. The condition of a fleece will depend on the management of the sheep throughout the year: any sickness, stress, or bad check in nutrition can cause a weakness in the wool being produced at that time resulting in an uneven or broken staple. A good fleece can be spoiled by stains or bad shearing with 'second cuts'; it can also be badly presented with daggles, straw, or other dirt, and carelessly rolled and tied. Apart from the breed and management throughout the year, the most important factors in producing top-quality fleeces are the organization of the shearing shed and the competence of shearers, fleece rollers, and packers. Dry, clean storage space is, of course, essential, since the wool may have to be stored for up to three months before it is collected by the Marketing Board. Packed in hessian wool sheets, it will keep indefinitely.

The art of shearing, by hand or machine, can be acquired only by experience and very hard work. Proper catching, lifting, and holding a sheep, together with an almost subconscious knowledge of anatomy are the foundations of good shearing. The rhythm and progressive movement of both shearer and sheep that lead to speed and accuracy, can come only with years of practice, but once the art of holding and handling is mastered, the actual shearing is not too difficult. A good teacher is worth many books; nevertheless Godfrey Bowen's well-illustrated little volume, *Wool Away*, is an excellent analysis of the skill of machine shearing. There are still a few old-timers who can teach hand shearing on the bench, but they are a dying race, so a beginner would be wise to hire a professional shearer the first year to teach him the skill. A shearing machine is an expensive item for a very small flock.

When I first came to hill farming, in the late 1950s, the tradition of communal gathering and shearing (mostly by hand) was strongly entrenched. Each farm had its own days for gathering, sorting, shearing, and turning out to the mountain, and its own circle of neighbours who automatically helped in all these tasks. No money changed hands and the greatest care was taken at every stage of the operation. The neighbours' wives came in to cook, serve, and wash up after the shearing lunch and supper and children helped with the paint pot, cut oil, and cider jug (generally called 'neck oil'). In the days of hand shearing, the gossip and story-telling went on endlessly to the accompaniment of the 'click, click, click' of the blades. Later, as machines took over, the intensity of work increased and talk was blotted out by the noise of the motors and shearing heads. Still, to this day, shearing can be a community effort in which the beginner can learn the craft step by step on his neighbours' farms while they bring their skills to his.

The sheep year

Until a few years ago, flock management, particularly on hill and marginal farms, was fettered by the rigours of winter and the farmer's limited ability to provide adequate winter fodder. It was generally accepted that ewes would start to sink in condition as the grass declined in the autumn, and that they would become relatively thin and weak as lambing time approached. The need to 'flush' the ewes and get them into rising condition before going to the ram was recognized as desirable but often difficult to achieve. Lambing percentages of 120 in downland flocks (65–70 per cent in mountain ewes) were accepted as adequate in bad years.

A greatly enlightened knowledge of the reproductive processes and their relationship to nutrition has resulted, in the past two decades, in a revolution in sheep management. It is now known that severe malnutrition, *at any time*, can damage a ewe's ovaries and reduce her potential fertility, and that stress and malnutrition in the weeks following tupping can cause reabsorption of the foetus and thus reduce the lambing percentages. This knowledge, skilfully broadcast by the Agricultural Advisory Service and the veterinary profession, is now common to most flockmasters who go to great trouble to keep nutrition at adequate levels *throughout the year* in order to achieve the optimum crop of healthy lambs. Hill flocks now aim at 120 per cent and often achieve 100 per cent, while downland sheep achieve 200 per cent or more.

A sound understanding of flock management throughout the year may be gleaned from *Sheep Farming Today* by J. F. H. Thomas, a pioneer in modern sheep husbandry.

Lambing

The most critical time in a shepherd's year is at lambing, particularly if he has a big flock to lamb in foul weather. The most vulnerable lambs are the first few (including the odd premature), which arrive over a period of several days, and the tail-enders when the shepherd is very tired and trying to get on with other pressing jobs. It is a great advantage if lambing can be compressed into as short a period as possible. This can be achieved by bringing the ewes to the ram in optimum condition and by stimulating their ovaries by the presence of rams in an adjoining field for a fortnight before tupping begins. Alternatively, an emasculated 'teaser' ram may be allowed to run with the flock prior to introducing the rams. This will tend to bring the ewes into season quickly. The ram may have 'raddle' (a red paint) smeared on his brisket – or a special harness holding a coloured marking block – to mark the ewes' rumps as he serves them. If you could be sure that a marked ewe had been properly served, you could then part her from the rest and make the ram's work easier (a ram will serve one ewe up to fifty times a day, ignoring others also in season). Personally I have found the system to be uncertain and time-consuming, although keen stock-breeders, using several colours, find it very helpful for recording matings and expected lambing dates.

The shepherd should be well prepared for lambing. A shoulder bag (the old wartime gas-mask haversack is ideal) containing obstetric jelly, iodine, towel, cotton wool, nylon cord, revival drench, a revival funnel and stomach tube, hypodermic syringe and antibiotics, marking pencil, and a length of sacking or blanket to wrap round a weak lamb, should be ready a week before the first lambs are expected.

Ewes in good condition and well exercised will generally lamb unaided. The first signs are the ewe separating herself from the flock and seeking a spot under a hedge, generally near the top of the field. She may paw the ground, turn in small circles, get up and lie down repeatedly. Later she will start to labour and the water bag will appear. If all is well the front feet and nose will be presented and the lamb will be born within an hour. If the lamb is not delivered within two or three hours the ewe should be examined in case help is needed to pull a large single from a maiden ewe or to correct a malpresentation.

When a ewe is assisted, the navel cord should not be broken for half a minute to allow the ewe's blood to be pumped into the lamb (in these thirty seconds the lamb's blood is increased by up to 25 per cent). If it has been necessary to insert the hand into the ewe to turn a lamb, there is always risk of dirt and infection; and here I would not hesitate to inject the ewe with a long-acting antibiotic.

In recent years many shepherds have taken to indoor management of the ewe flock. The same arguments, pro and con, apply as with cattle, with the added problem that, if the weather is mild, ewes can become overheated indoors. There is no doubt that most casualties at lambing are due to exposure and hypothermia, and a sheep-sick lambing meadow near the farmhouse is every bit as infectious as a shed, which at least provides shelter from driving rain or snow. Ewes kept indoors before lambing cannot get the necessary exercise, and malpresentations seem more common indoors than out.

A novice shepherd would be well advised to attend a one-day course on 'Lambing Techniques' run in many districts jointly by the Agricultural Training Board and local vets. Using the 'Phantom Ewe' equipment, it is possible to learn how to correct and deliver a malpresented lamb under expert supervision and at the same time to discuss and resolve the many questions that assail a beginner.

Feeding

Provided the farm is not overstocked, sheep will thrive on grass alone for seven or eight months of the year; indeed they show little interest in hay or roots if sufficient clean grass is available. Sheep-tainted pastures, however, are spurned, even if there appears to be plenty of grass.

As soil temperatures drop and the ground becomes sodden

in November and December, the feeding value of grass drops considerably, and recently tupped ewes may sink in condition. This can be a critical time for the new foetus which may be reabsorbed if the ewe comes under nutritional stress. It is most important to maintain body condition at this time; supplementary feeding on foggage, autumn-sown oats or rye, rape, and turnips or swedes may be necessary. It is unlikely that sheep will eat hay before Christmas unless there is snow and hard frost.

In the depths of winter, from Christmas until lambing, grass scarcely grows at all, and what there is has little value. This is the time when the foetus grows rapidly, putting a heavy and increasing demand on the ewe. Well-made, soft hay should be fed ad lib, gradually being replaced by more concentrated food as the foetus grows and reduces the rumen space in the ewe's body cavity. Care must be taken not to overfeed in-lamb ewes. Large single lambs can result in difficult deliveries. I have found that sugar-beet nuts (building up to 225 g. a day) or rolled oats (up to 115 g.) is enough for my Black Welsh Mountain ewes: heavier breeds would require more. I feed about 30 g. of seaweed meal per ewe per day as a mineral supplement. Some succulent food is also required: I feed about $1\frac{1}{2}$ kg. of swedes per ewe per day from Christmas to mid-February and a similar quantity of mangolds from then until after lambing.

By the end of the previous September, I will have laid up a paddock to be ready for the ewes and lambs in spring, and turn them into it the day after they lamb. This clean, fresh sward stimulates the milk flow and gets the lambs off to a good start. I seldom have enough of this good grazing to see the flock through to the summer, and generally have to revert to mangolds in May. At this time, my grazing is further reduced by the need to lay up fields for hay, and the grass crisis continues until the hay is cleared and the aftermath has grown – probably by mid- or late July.

Until a few years ago, lambs had to be fat in order to grade and sell; today a fat lamb is penalized in the market. Lean carcases that would previously have failed to grade are now the butcher's ideal, and supplementary fattening crops such as rape and turnips are seldom required. Nevertheless, it is always useful to have an autumn forage crop to maintain the ewes' condition until Christmas.

Well-grown ewe lambs may be tupped at seven to eight months old, although many farmers consider it wise to hold them back until the following year. If they are put in lamb early, they require extra feeding to enable them to grow as well as produce a lamb. If they are not tupped they can be stored on a lower plane of nutrition for their first winter, and will become stronger ewes by the following autumn.

The ram

A fertile ram can cope with forty to fifty ewes. Bearing in mind the (remote) chance that any one ram may be sterile, I would run two rams with a flock of fifty, three with a hundred, and so on. In a small flock of ten to twenty, I would take a chance on one only. In a flock of fewer than ten ewes it might be better to borrow a well-grown ram lamb or run the ewes with a neighbour's flock. A ram's working season is only six to eight weeks long. In this period he works very hard and will probably flag if he does not start off in prime condition. For the other ten months he is a passenger and may be rather a nuisance. There is often a tendency to neglect him, only to find he is off his feet and in poor shape when next needed. It is important to see that he is worm-free, sound of foot, and not bothered by fly-strike (particularly around the horn roots) by early September, and that he is on good keep from then until tupping time.

My ram runs with the flock from the beginning of tupping time (late October) until the lambs are weaned (late August), when he is parted to avoid too early contact with the ewes or with the ewe lambs. I usually arrange for him to join my neighbour's rams so that he is not alone and tempted to break through the strongest hedge!

Sheep shelters

The vast majority of sheep are managed entirely out of doors, hardiness being an essential quality of the species. But, like other livestock, they lose condition under stress of foul weather and inadequate feed. New-born lambs are particularly vulnerable to cold, wet winds. The provision of limited shelter, particularly at lambing time, is both humane and good economy. In recent years, as nutrition of the pregnant ewe and unborn lamb became better understood, hay, silage, and concentrate feeding throughout the winter has increased, together with the provision of shelter. The next step was the provision of covered accommodation for lambing (either on straw or slatted floors). These arrangements, plus electric light, made lambing much easier for the shepherd as well as protecting new-born lambs from the weather. A few flockmasters now bring their ewes indoors for the whole winter, creating, as nearly as possible, a controlled environment. By this device it appears possible to increase the number of breeding ewes (on which a subsidy is drawn) that a farm can hold, to attain a higher lambing percentage of earlier-born lambs, and to feed 'scientifically' the exact amount the ewes require. But there are snags: without sufficient exercise, ewes tend to mispresent their lambs; crowded together they overheat and may need shearing in February; diseases can become established in the buildings and bedding and respiratory ailments may be encountered; and when the increased numbers of sheep are turned out for the summer they may well be too thick on the ground. Although this system may seem to make economic sense, in the long term it can prove disastrous, and not worth the erection of a specially designed building.

For some years, when I managed a medium-sized hill flock, I provided a slatted-floor lean-to shelter for my ewe lambs in their first winter instead of sending them down-country 'on-tack'. The lambs ran out onto a ten-acre field; they were fed indoors on meadow hay and sugar-beet nuts. After a few weeks the field became little more than a muddy exercise yard, and, although I saved money on 'tack', the lambs did not grow as well as those sent away to a lowland farm to winter. From time to time I have allowed my dozen or so ewes access to an open shed containing a hay rack in bad weather. Most of them lie indoors all night and go out by day. It is always wise to have a few pens of hurdles or straw bales at lambing time for ewes with weak or fostered lambs.

Observation: the key to management

The old saying 'It's the master's eye that makes the horse grow fat' is even more applicable to sheep. Not only are they prone to a host of ailments and accidents, but, if not helped promptly, seem to acquire a determination to die. A few visits by the vet can wipe out the profit from a small flock; prevention must be the watchword. Being flock animals, they are often kept too thick on the ground which can become sheep-sick and spread infection. 'A sheep's worst enemy is another sheep' is also a truism worth serious thought. Time spent observing sheep; how they stand, walk, run, lie down, suckle, and bleat, is never time wasted provided the shepherd interprets and acts upon what he sees.

Goats

Goats are kept primarily for their milk, either as a substitute for the smallholder's house cow or for the commercial sale of milk, butter, cheese, and yoghurt. Very small holdings, which could not support a cow, might still manage of pair of goats. Goat meat from surplus kids is a useful supplement to the family diet, and may sometimes find a market with local poulterers, but is of relatively insignificant commercial importance in Britain.

In spite of physiological similarities, a goat is very different from a sheep. It is primarily a browser rather than a grazer, it carries an oilless hairy coat instead of a greasy woollen fleece and cannot tolerate cold, wet, windy weather; it is more nimble than the most active sheep and is inclined to produce a much higher daily milk yield over a far longer lactation. It adapts to indoor management or tethering (the latter not universally recommended), and becomes very domesticated and affectionate.

Goats are resistant to TB and brucellosis and their milk forms a very soft, digestible curd, highly suitable for children. The fat globules are very small, and the cream does not rise quickly. Under hygienic dairy conditions there is no 'goaty' taste to the milk or other products, although the males produce a very strong smell in the rutting season.

As they do not tolerate heavy rain, goats are not managed like sheep in Britain. They are generally brought indoors at night and kept in during bad weather. As they are extremely active and curious, they are not easy to restrain, and can do great damage to hedgerow, orchard, or garden if not kept under strict control.

Contrary to much misinformed opinion, goats cannot live on a mixture of air and waste paper; indeed they require a carefully balanced diet to maintain health and the high milk production of which they are capable.

Like sheep, however, goats are 'flock' animals, and do not thrive in isolation. It is therefore wise to keep two or more on a holding.

Breeding

The various breeds of goat and their characteristics are well described in L. V. Hetherington's excellent book *Home Goat*

Keeping. The British Goat Society in Suffolk and affiliated clubs throughout the country provide a readily available source of advice and information on every aspect of goat breeding and management. A goat owner is strongly advised to join the nearest goat club.

Unless he plans to run a dairy flock, a smallholder would probably be ill-advised to keep a male. During the five-month rutting season a billy develops a most unpleasant, pervasive odour, must be specially housed away from the females, and requires feeding throughout the year although he produces no milk. A smallholder can always take his goat to a well-bred male of a larger flock.

Goatlings are seldom mated until their second autumn, and kid when two years old. Twins and triplets are very common, and are normally hand reared from the fourth day when the colostrum has disappeared. Most goats have long lactations and do not need to kid every year. As milk is the principal aim and meat of minor importance, the production and rearing of kids is not very profitable. It is generally advantageous to make the most of the long lactation, keep two goats, and kid them in alternate years.

Kidding

The symptoms and sequence of kidding, which generally takes place indoors, are very similar to lambing, and similar hygiene and assistance are applicable. Rather more human intervention is normal, particularly in bottle-feeding of colostrum and subsequent rearing of the kids if they are retained.

Feeding

A successful lactation depends on correct feeding, and the first requirement is bulk. High-quality clover, lucerne, or meadow hay forms the basis of goat nutrition, and can be fed with little waste in racks in the goat house. Pea haulms and bean stalks are also good sources of bulk.

Succulent foods include grass and leaves of trees and bushes, outer leaves and stalks of vegetables, root vegetables (mangolds are NOT suitable for males), fresh hedge trimmings, and windfall fruit. Sugar-beet pulp (soaked for milkers and in-kid goats) is a valuable addition to the winter feed.

Cereals and bran with high-protein oil cake (groundnut, linseed, or soya bean) make up the concentrate portion of the ration, rolled oats and bran providing the main ingredients. A little seaweed meal with each feed before milking or fed ad lib from a hopper provides mineral trace elements.

Like sheep and cattle, goats are best fed to keep their body condition right throughout gestation and lactation. As goats spend so much more time indoors than sheep, it is easier to control the diet and anticipate the seasonal changes in grazing or browsing material.

Housing

Goats may be successfully housed in converted sheds or farm buildings, or special accommodation may be provided. The most important requirements are that it should be draught-free, reasonably insulated against cold and condensation, have a well-drained floor, and be light and well ventilated. Ease of access, a convenient water supply, electric lighting, and space for storage of feeding-stuffs and equipment should all be carefully considered. It is also important to have a convenient milking place with a stand to raise the goat to the right height.

Control

Being such agile browsers, goats can do great damage to hedgerow, orchard, or garden unless well controlled. The alternatives are to confine them to house and concreted yard, tether them, or ensure that your fences are completely goat-proof and let them run free. A concrete yard is convenient and secure, but means that all succulent food must be carried to it, and that exercise, an important factor in maintaining health, is limited. Tethered goats can graze or browse road verges, orchards and other food sources where it would be impossible to turn them loose; but this involves frequent moves, the carrying of water, and prevents the goats from finding shelter if the weather turns foul. With good fences and free access to shelter, goats can look after themselves with minimum trouble to their keeper, but don't underestimate the cost in time and materials of making fences goat-proof. Goats can be trained to respect electric fences, either multi-strand wire or flexinet mesh. This system is relatively cheap, quick to erect and move, and effective so long as it is carefully erected, regularly checked, and well maintained. If the circuit is broken or shorted out, a clever goat will soon find out. The plastic conductors gradually deteriorate in sunlight, and badgers and rabbits are liable to chew through the droppers and bottom strand.

Probably the ideal compromise is a combination of the three methods: free range on secure fields, tethering elsewhere when convenient, and a secure yard for when you cannot keep a close eye on your animals.

Expertly managed, goats can be a very worthwhile enterprise on a smallholding but will prove disappointing and troublesome if taken on without knowledge and careful planning.

Pigs

Modern pig-keeping is very big business with keen international competition. Commercially, a smallholder is in a poor position compared with the big operators and marketing groups who, through economy of scale, can make useful profits from narrow margins. Yet there are still ways in which you can keep a few pigs to advantage, if you know your job and are prepared to give enough time and energy to their management.

The most obvious benefit is meat for your own table, which you can produce considerably cheaper than you can buy it from the retail butcher. In doing this, you will be using up surplus skim milk, whey, and waste products from other farm enterprises. A less easily measured profit will accrue from the pig manure – a very valuable addition to the compost heap. If you breed from a sow, there will be a few weaners to sell in the market (or you might join a weaner group), or you may sell breeding stock of one of the rarer breeds to other smallholders. A final bonus may lie in reclamation of small parcels of scrub land by rooting pigs.

Pig-keeping can embrace the whole range of enterprises from breeding to bacon-curing, or it may be confined to any one of the several stages. You may buy a fat pig to kill and cure, a weaner or porker to fatten to bacon weight, an in-pig sow or gilt, or weaner gilts to become breeding stock. You may or may not decide to keep a boar. This will depend on the availability and location of a stud boar. Three or four sows can justify keeping a boar, and you might make him available to neighbours. Whichever enterprise you choose, you will do well to think it through and plan your accommodation, timing, feeding, and marketing before you buy any pigs.

The various breeds and their characteristics are clearly described in Thear's *Complete Book of Raising Livestock and Poultry*. Unless you plan to go into a rare breed, you would be wise to study your local markets and the pig-keeping of your neighbours in order to select the most suitable breed for your farm.

Breeding

A well-reared gilt will be ready to breed at six to eight months of age. The signs of heat are a slight swelling and pinkening of the vulva and a tendency to mount other sows. If a boar is in close proximity, the heat symptoms will be readily evident, and she will stand to the boar. If you haven't got a boar you will have to take your gilt elsewhere for service, and may have to leave her there to ensure a successful mating. Farrowing takes place 115 days after mating.

The sow will not come 'brimming' while piglets are still sucking, but will do so four days after they are weaned. It is normal practice to wean at eight weeks and take the sow to the boar four to five days later. By this means it is possible to get exactly two litters in one year. Early or very early weaning is also practised, but this requires expert stockmanship, more concentrated feeding of the piglets, and puts a heavier strain on the sow, who is expected to bear three or more litters a year. On balance I believe that, in the long term, what you gain on the swings, you lose on the roundabout, and I prefer to stick to eight-week weaning. It is worth trying to time the first farrowing for March/April or September/October, particularly when on outdoor management, so that the new-born piglets avoid the worst of the winter weather.

Housing

Pigs are naturally beasts of the forest, and find a well-balanced diet on the ground or in the topsoil.

They need shelter from wind and weather, but can tolerate very hard and simple conditions. In some intensive systems they are cosseted in a controlled environment for maximum production from minimum foods, but this requires great skill in feeding and management.

Simple outdoor shelters can be made from straw bales and

corrugated iron. With a deep, dry bed of straw and shelter from the prevailing wind, sows will make themselves very comfortable and farrow successfully in such houses. To avoid destruction of the shelter, the straw bales should be wrapped in rabbit wire and the whole well staked to the ground; a crush bar to protect newly farrowed piglets from their mother's weight should extend down one side of the shelter. For outdoor management, fences must be secure and regularly inspected, and pigs' noses ringed if you do not want them to root up the fields.

A traditional pigsty with weaner, sow, and boar quarters, and exercise yard and feeding facilities is a satisfactory way of controlling pigs for breeding, or fattening. A dry, insulated floor, protection from draughts and rain, good ventilation, plenty of bedding and access to fresh water are the bare essentials. This system is convenient for the pig-keeper, pigs will grow and fatten well; but it does not produce quite such hardy, healthy pigs as outdoor management.

Kennels, cubicles, and more sophisticated housing with controlled environment and slatted-floor dunging areas are described in *The Complete Book of Raising Livestock and Poultry*. In the more intensive systems, crates, sweat-boxes, and very restricted fetters are used, but these devices are part of factory farming, and not, in my opinion, suitable for smallholders (or pigs!).

Feeding

Unlike ruminants and horses, pigs have simple alimentary canals (similar to human beings) and are unable to digest bulky, fibrous foods such as hay and straw. They require more concentrated and easily digestible proteins, carbohydrates, and succulents: potatoes should be cooked. Pigs housed indoors and unable to root in the ground tend to become deficient in iron and vitamins, and piglets develop anaemia. To offset this they should be offered turf and compost in their yard, plus a little seaweed meal in their food to provide trace-element minerals.

A piglet's ration should be high in protein (up to 18 per cent for growth) gradually giving way to more carbohydrates after eight weeks for fattening. The protein factor can come from fish-meal, soya meal, whey, or skim milk; the carbohydrate from barley meal. Green food in the form of vegetable leaves and stalks and grass is an important source of vitamins. Pig nuts, containing the required balance of protein, carbohydrate, and minerals, may be bought in, although I prefer to mix my own. Because of the danger of salmonella, feeding pigs on swill is subject to strict regulations. Boilers for sterilizing swill are expensive to buy and operate.

Table 6 shows examples of mixtures suitable for breeding sows and weaners, growers, and fattening pigs.

Table 6. Percentages of ingredients for mixtures suitable for breeding sows and weaners, growers, and fattening pigs

Meal mixture	Nursing sows and weaners	Growing pigs 3–5 months	Over 5 months
White fish-meal	7	4	0
Bean meal	7	9	13
Barley meal	25	30	35
Flaked maize	25	25	20
Wheatings	35	30	30
Minerals	1	2	2
Starch equivalent	68	67	67
Protein equivalent	13	11.5	10

Source: Watson and More, p. 759

A pig up to 25 kg. live weight requires a daily ration of approximately 5 per cent of its body weight, gradually dropping to 3 per cent when its live weight reaches 115 kg.

Killing and preserving

Pig-killing on the farm was formerly something of a social occasion. Neighbours gathered (including one licensed to slaughter pigs) and a gruesome ritual was performed. The death was not always as humane as it might have been. I prefer to take my pig to the nearest Fatstock Marketing Corporation (FMC) slaughterhouse where it is despatched humanely and cut up for a very modest fee.

Porkers can be deep frozen immediately, but if you intend to salt a baconer, it is wise to go over the sequence and prepare the site, equipment, and materials in advance.

There are many cures, including smoking, as well as recipes for faggots, brawn, trotters, small meat, and sausages which are beyond the scope of my small book.

Chickens

Poultry, like pig-keeping, is now multinational big business, and it is well-nigh impossible for a smallholder to compete in the commercial egg and table-bird markets, even if you adopt factory-farming methods. On the other hand you may find a local market for free-range eggs, layers, or table birds, which you can sell at a premium and clear a small profit. With a few hens, ducks, or geese, fed on household and farm scraps, you can certainly provide top-quality eggs and the odd table bird for your family much cheaper than buying them in the supermarket.

Poultry enterprises fall into three distinct categories: egg production, breeding stock, and table birds. A self-sufficient smallholder might wish to do the lot on a very small scale. Others, keener on showing a profit, would tend to concentrate on one aspect only.

Egg production

Free-range egg production is a matter of acquiring the right breed and strain of birds at the right time and feeding them with the right food in a suitable environment. Details of breeds, feeding, and housing are discussed later in this chapter.

A good strain of laying bird, well fed on free range, can be expected to lay about 220 eggs in its first year and to drop by about 25 per cent each year thereafter. In my view it is worth keeping birds on for a second year's laying, but after that (that is when two and a half years old) they are best fattened and killed.

You have five options for acquiring your laying birds: fertile eggs, bought in or from your own birds and hatched in an incubator or under broody hens; day-old chicks, sexed or as they come; growing pullets; point-of-lay birds; or 18-month-old battery birds.

Hatching eggs from a reputable breeder needs the smallest initial outlay, but involves the risk of failure, the trouble of hatching, feeding to point of lay and coping with the 50 per cent which will be cockerels. Day-old chicks are also cheap to buy and will be well acclimatized to their new home by the time they are at point of lay. They may or may not be sexed before despatch. Sexed growers at ten to twelve weeks are more expensive to buy. Point-of-lay pullets, at five to six months, are much costlier, but by far the easiest and safest way to start a laying flock. Probably the cheapest start of all can be made by buying in battery hens, culled after their first year of pressurized laying. They look dreadful, have no idea how to cope with free-range conditions, are not hardy, and may be diseased. But they soon learn and toughen up, and with good luck they may provide eggs for the family very cheaply indeed. I would not risk this method except on a very small scale.

My personal choice is to buy Rhode Island Red x Light Sussex day-old pullets, hatched in March or April, rear them outdoors to point of lay by the end of September, keep them free range for two laying seasons, and fatten them after their summer moult to kill at about twenty-eight months of age.

Breeding stock

There is a small but steady demand for hardy laying birds of the old-fashioned pure breeds and first crosses. The big hatcheries, providing birds for batteries and broiler houses, concentrate on sophisticated hybrids developed for extremely high production and short lives in artificial conditions. If you have a delight in poultry and a well-set-up system of incubator, brooder, and free-range arks you should have no trouble in selling day-olds or point-of-lay pullets surplus to your own needs. But if you take on such an enterprise you cannot escape the need to rear and dispose of the cockerels. A few of the best, if pure bred, you may sell for breeding, but the majority will have to be fattened, killed, plucked, dressed, and marketed – a difficult and laborious business unless you are specifically geared to it.

Table birds

You are unlikely to make table-bird production your main poultry venture unless you have a guaranteed retail market in a hotel or health-food restaurant. If you do plan to fatten any quantity of table birds, they should be of a dual-purpose or heavy breed or cross with acceptable carcase and flesh qualities.

With sex-linked crosses (such as Rhode Island Red x Light Sussex, where the pullets are buff-coloured and the cocks white) the cockerels can be reared separately from hatching. It is more usual, however, to sex and separate chicks at ten to twelve weeks. The cockerels should then be reared on a high nutritional plane to be ready to kill before they reach sexual maturity at eighteen to twenty-four weeks. Failure to do this can lead to fighting and rapid loss in condition. There are two alternatives, which I personally dislike: to caponize, or fatten in individual crates or cages. In either case the birds can be taken on to heavier weights. Surgical castration is now illegal; caponization is achieved by implanting a hormone pellet behind the bird's ear. Fattening crates are laborious and to my mind inhumane, and it is not always easy to maintain the birds' appetites.

Cull layers at two and a half years of age can be quickly fattened and killed before starting to lay in their third season. Inevitably some of them will have full ovaries and would go on laying for a year or two more, but on balance it is not economic to keep birds for a third season.

Killing, plucking, and dressing is expertly described in Geoffrey Eley's book *Home Poultry Keeping*.

Breeds

The ideal bird (not yet achieved) would be hardy, quick growing, of good size, easily maturing, a prolific layer, not too prone to go broody (yet there are times when a few broodies are useful!), well covered with white flesh, an excellent converter of food, and long living.

Pure poultry breeds can be divided into three main sections: light, mainly for egg production typified in the White Leghorn; heavy, mainly for fattening, such as the Light Sussex; and dual purpose which include White Wyandotte and Rhode Island Red. First crosses between such pure breeds produce excellent dual-purpose birds with hybrid vigour and, in the case of Rhode Island Red x Light Sussex, sex linkage in the chicks. Unless they are back-crossed to one of their pure-bred ancestors, first-cross hybrids will not breed true and are therefore useless as breeding stock.

Modern hatcheries produce very sophisticated hybrids for egg or broiler production, and it is no longer easy to find sources of pure breeds or even first crosses. A smallholder might do well to buy point-of-lay hybrids from a commercial breeder for egg production.

Fanciers, who breed for showing, do keep pure breeds, but these often lose their laying or table potential in favour of colour and show points.

Hatching

Hatching chicks from fertile eggs in an incubator is straightforward if plans are carefully laid, equipment well maintained, and attention paid to detail. You should start with the desired end product and work backwards; only then is it possible to begin collecting eggs at the right moment, set them in the incubator, and prepare the brooder, hay box, and arks in timely order. Assuming that we want to produce 50 point-of-lay pullets by 1 October (at five months of age), they will need to hatch about 1 May, be set in the incubator on 9 April, which should have been lit on the 5th or 6th and brought up to a steady, correct heat – 39 °C one inch above the top of the eggs before they are put in. To produce 50 pullets at point of lay, I would suggest 150 eggs be set. If my birds were laying on average 35 a day and I needed five a day in the kitchen, it would theoretically take five days to collect 150. As a few would be rejected (too small or too large, soft or misshapen shell) I would start to collect on 2 April, a full week before they are due to be set. I do not recommend keeping eggs for more than ten days, as their fertility tends to drop. If they are kept in a humid atmosphere or buried in bran, they are less likely to dry out.

In the incubator, the eggs need turning twice daily, the water tins must be regularly topped up, the incubator wick trimmed, and the temperature regulated; if it rises above 40.5 °C for more than a few hours, the embryos will be cooked; if it drops for long below 37.8 °C (a lesser evil) they may die of cold. Eggs that fail the candle test (in which a light is shone through the egg to show if the embryo is developing) at ten

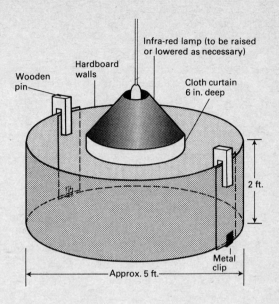

Fig. 16. A simple brooder

This brooder is designed for up to 50 day-old chicks and is made from one sheet of 8 ft. x 4 ft. hardboard halved lengthwise. A fine-mesh wire lid may be necessary to exclude cats or rats.

and sixteen days should be removed lest they crack and pollute the rest.

Small numbers of birds may be hatched under broody hens. With modern breeds, broodiness is not always strong and some birds give up after a day or two. It is therefore wise to let a bird sit on china eggs for three or four days before entrusting a clutch to her. A small coop and wire run should be provided in a secluded spot and the bird encouraged daily to get off to dung, eat, and drink; otherwise she should be left strictly alone.

Brooding

Chicks need not be fed or removed from the incubator for forty-eight hours after hatching. They should then be moved to a draught- and rat-proof brooder with artificial heat (see Figure 16).

Depending on the weather, the heat can be progressively withdrawn, and the chicks transferred to a movable hay-box house. This is normally done at about three weeks, although I have had chicks successfully out on short grass when four days old.

Growers

The hay box should be moved daily and the chicks shut inside at night to keep them warm. As they grow older, top ventilation is increased and the chicks are allowed to go out on free range by day. At ten to twelve weeks they are ready to be sexed and separated. The pullets can then go into uninsulated night arks and the cockerels to fattening quarters. Running free range by day, the pullets grow well and are extremely tough by the time they are ready to move to the laying quarters.

The laying flock

Depending on the buildings and space available, there are several methods of housing a laying flock. Excluding battery houses and intensive deep litter, which amount to factory farming, systems vary from semi-intensive, movable arks with limited access to grass to a fixed laying house with complete free range over fields. An excellent system has been devised by Lady Eve Balfour and described in *Keeping Poultry on a Small Scale*: it comprises an outdoor strawed yard with controlled access onto grass leys and a simple laying and perching shed. It is not, however, a very suitable system for high-rainfall areas where the straw yards and leys could become badly poached. Other methods for small and medium-sized flocks have been developed by Jim Worthington and described in his book, *Natural Poultry Keeping.*

Feeding

Newly hatched chicks require no food for the first forty-eight hours of their lives, living off the remainder of their yolk sacs. They should be introduced to finely kibbled mixed corn (or chick crumbs) spread on a flat board over which they can run. I start at about 14 g. a day in three feeds, building up to 57 g. at four weeks. The rate of buildup is based on what they clear completely. With each feed I scatter a little fine flint grit, and always ensure that there is plenty of clean water in a safe, shallow drinker. As soon as weather permits, the chicks are transferred to a hay-box house and allowed access to *short* grass (long grass can ball up and block the crop). At this point the feed is offered in narrow troughs or on dry ground twice a day: grower's crumbs in the morning (with a higher protein content) and mixed corn in the evening. Alternatives are to mix your own mash or to offer protein meal ad lib in a hopper: the latter only if the birds are on plenty of grass. A self-feed hopper of medium flint grit is kept filled, and as soon as they are big enough to deter cats and carrion crows (about ten to twelve weeks), they are allowed to free-range over grass by day. The foundations of healthy birds are built from *grain*, *grass*, *greens*, and *grit*. On good grass the growers will pick up a lot of protein in the form of grubs and worms, and will tend to balance their own ration. From ten weeks, oyster shell is available as well as flint grit to ensure that their eggs have strong shells.

From point of lay, a high-protein ration must be offered, as the birds come into heavy production. From this point I offer skim milk or whey as an additional source of protein.

Birds in moult tend to become overfat if fed on a layer's diet. The carbohydrate (cereal) portion should be replaced by bran. Access to grass and exercise are particularly important.

When fattening fully grown birds, the protein content is reduced in favour of carbohydrates (such as barley and maize meal). To fatten young cockerels quickly from ten weeks, the protein must be maintained to promote growth and the carbohydrates increased for rapid fattening. Exercise should be somewhat restricted and fighting prevented. A very sound book (now out of print and available only from libraries) is Beatrice Malcolm's *Successful Poultry Farming*. In it are the proportions of home-mixed feeds for every class of bird.

Caponized cockerels grow larger and fatten more easily than 'entire' birds. As stated earlier, surgical castration is now illegal. Hormone insertion (either via pellet or in the food) is still legal in Britain although forbidden in Australia and the USA.

The smallholder with only a few hens can provide much of their feed from farm and kitchen waste. Vegetable remains, chat potatoes (too small for human use), waste garden greens, skim milk, and whey provide a great deal of protein. Threshing-floor sweepings including weed seeds may provide much of the grain. If he can guarantee his birds grain, grass, greens, and grit, they should remain healthy, and economic, even if they do not break laying records.

Rabbits, ducks, geese, and turkeys

I have not kept these animals, so I recommend Thear's excellent book on livestock and poultry already referred to.

Horses

A great many aspiring smallholders dream of a return to the horse-power age which has almost disappeared from British farming since the end of World War II. With the decline of the working horse the human skills have been all but lost, and today there are very few young or middle-aged horsemen capable of harnessing and handling a team, much less turning a straight furrow. To work well, horses must be fit and worked regularly, day in, day out. To keep horses fit involves long hours and strict discipline and is labour intensive for both man and beast.

I can name a few smallholders who bought working horses, harness, and tackle only to find they had not the time to use them fully. I am one myself. I can name others who bought horses without the slightest idea how to handle them and found them not only disappointing, but a disaster.

Nevertheless, I believe the horse is coming back: first on large farms where certain jobs such as hoeing, harrowing, and carting can be done better and more economically by horses than by buying extra tractors; and eventually, as machinery and fuel become even more costly, horse power will spread down to smaller and smaller farms.

Having handled horses since childhood, and trained both horses and bullocks to farm work during the past twenty years, I feel impelled to urge anyone contemplating the use of horses on his farm to learn what is involved before committing himself. Practical courses are available and any amount of excellent literature exists in libraries. Working horses can be bought, some of them trained and worth the money. Good harness is terribly expensive and implements few and far between. Perhaps the most elusive commodity of all in this hectic age is the time required to look after, school, and work a horse.

Stables and harness room

Horses and ponies need shelter from wind and rain, and, if working regularly, need proper stabling. A lean-to shed against the prevailing wind may be provided in the paddock.

It should have a concrete floor, and sufficient space and a wide enough entrance to allow easy exit should a kicking match take place.

Stabling must provide a dry bed, protection against draughts, water, mangers, and racks. Horses may be tethered in stalls, or provided with loose boxes. Unlike cattle, horses need separate boxes to avoid injury from kicking. Stalls may be permanent fixtures or temporary hanging 'kicking poles'.

A harness room is essential for a farmer using horses. Badly stored and maintained leather becomes brittle and weak, stitching rots, and buckles corrode. Repairs and renewals are very expensive. The room should be dry and ventilated. Pegs, brackets, or racks are required and should be round-topped to avoid sharp bends in leather straps. Cleaning hooks, a saddle horse, a cupboard for cleaning materials, water, and electric light complete the ideal tack room.

Bees

I have kept bees for many years, and am sure they should find a place on every smallholding. In good years they will provide enough honey for the family with a surplus to sell. In bad years they will still be of immense benefit to orchard and garden by spreading pollen and fertilizing fruit and vegetables.

The dedicated, serious beekeeper can spend a great deal of his time about his hives. The busy mixed farmer will probably say 'leave them alone and they'll come home'. From my six hives I occasionally take as much as 300 lb. of honey; on average the yield is little more than 100 lb. Properly extracted and bottled, honey fetches a good price, but bee equipment is extremely expensive. I am not a dedicated beekeeper, and go into my hives as seldom as possible. In April or May, when the bees are working hard, I open the hives to ensure that the queens are laying, remove the winter mouse-guards and pinch out any queen cells. If I find a weak colony with little honey, I feed some sugar syrup. During May, June, and July I attempt to take and hive any swarms that occur. In August or early September I take and separate the honey, and in October I examine all the hives for residual honey (feeding sugar syrup to make up any deficiency) and replace the mouse-guards at the entrances. If there is heavy snow during the winter, I clean the entrances to ensure ventilation.

Beekeepers' Associations cover the whole country and are most helpful brotherhoods. Practical courses are available at agricultural and further education colleges. Volumes have been written on this complex and fascinating subject, which could not possibly be covered in a book of this size.

Health and hygiene in livestock

Maintaining good health in animals should be a positive process. True health, built up throughout the life of the animal, carries with it hardiness and the power to resist and overcome disease and is based on good breeding, sound nutrition, and the right environment. In theory a good farmer should be able to provide all these conditions in full measure; indeed some carry the matter to extremes of mechanically controlled environment, and theoretically perfect nutrition. But practical farming is always a compromise in which allowances must be made for many unforeseen factors. The quality of feeding-stuffs may be difficult to assess, the weather unpredictable, markets disappointing; the farmer may find himself short of feed, overstocked, overworked.

Even if his home-bred, acclimatized stock is healthy, he may buy in a diseased beast. Accidents occur on the best-run farms. The farmer's job is to build good health, to spot the first signs of disease, to find and remove the cause and then to treat and overcome the symptoms.

Daily checking and handling of livestock will give early warning that something is wrong; at this stage it is usually possible to prevent the condition from becoming serious. Too

Table 7. Causes, symptoms, and treatment of diseases

Disease	Probable causes	Symptoms		Treatment	
		Early	Late	Early	Late
CATTLE Mastitis in milking cows	Poor hygiene, cow not being milked out. Cow under production stress and high concentrate feeding. Physical injury to udder.	Hard, hot quarter. Lumpy teat. Clots in milk. Cow tender to milking. Milk slow to pass through filter.	Swollen, hard, hot quarter. Distended teat. Pussy discharge instead of milk. Extreme tenderness. Loss of appetite. Cow runs a temperature.	Starve for 48 hours. Drench or feed with garlic twice daily. Strip out three times a day. Cold water hose on udder for 10 minutes, three times a day. Resume feeding on third day on succulents and hay only. Reintroduce concentrates gradually. Watch carefully until completely cured.	(If not spotted in time) strip out three times daily. Antibiotics. Reduce concentrates and replace with succulents. If in doubt, call vet.
Summer mastitis in dry cows or nurse cows at grass	Cow not properly dried off. Physical injury to udder. Flies. Cow produces too much milk for calf.	Cow kicks at calf. Hot, hard quarter. Distended teat(s). Calf tends to scour.	Swollen quarter. Distended teat(s). Flies on teats. Pussy discharge. Loss of appetite.		
Milk fever in heavy milking cows	Non-availability of calcium in bloodstream.	Cow looks weak and unbalanced at or within 48 hours of calving.	Cow down and unable to get up; may go into a coma.	Subcutaneous drip 2 litres calcium boroglucinate: do not strip. Alternatively, inflate udder with air pump after stripping.	Call vet. Intravenous injection calcium boroglucinate.
Coughing in calves (spring and early summer)	Poor, dusty hay. Lack of fresh air.	Slight dry cough.	Deep bubbly cough.	Feed only best hay. Increase ventilation. Turn out by day.	Call vet.

Table 7 cont.

Disease	Probable causes	Symptoms		Treatment	
		Early	Late	Early	Late
Lung worm or husk	Lung worms picked up in overstocked pastures in autumn.	Coughing in autumn.	Heavy coughing and choking. Worms appear at mouth and nostrils.	Bring indoors. Fast for 2 days, drenching with garlic 4 times a day. Then on dry food and drench with garlic daily for 7 days. Preventive: Drench with 'Dictal' before turning out in spring.	Almost impossible to cure in late stages. Preventive: Keep calves on clean pastures in autumn.
New Forest disease (infectious keratitis)	Fly-borne organism 'Moraxella'.	Eye half closed and weeping.	Blue-grey film over eye leading to ulcer and permanent blindness.	Dilute lemon juice or cucumber juice, or antibiotic and anaesthetic powder. Isolate bearer.	Continue treatment.
SHEEP Gid	Brain cyst caused by tapeworm which has been hosted by dog or fox.	Sheep sometimes walks in small circles.	Sheep grazes less and less, always walks in small circles. Soft spot on temple.	Cyst may be surgically removed by vet. Bury all carcases. Keep all dogs wormed.	Bury all carcases. Keep all dogs wormed.
Fluke	Invasion of liver by fluke. Wet areas where fluke snails thrive.	Sheep lethargic and in sinking condition. Livers of slaughtered lambs reported to show signs of fluke.	Puffy swelling under lower jaw. Sheep in very poor condition and lethargic.	Fluke and worm drench. Drain wet sites where snails breed. Copper sulphate on wet sites to kill snails or introduce ducks to eat snails.	Generally no effective treatment of sheep. Deal with cause.

Table 7 *cont.*

Disease	Probable causes	Symptoms		Treatment	
		Early	Late	Early	Late
Worms	Overstocking. Dirty pasture.	Sheep do not thrive. Scouring. Worms in dung.	Sheep sink in condition. Bad scouring.	Move sheep regularly to well-rested pastures. Ensure lambs are well fed (plenty of ewe's milk). Worm drench at start of summer grazing rotation.	Repeat worm drench. Keep on clean pastures. Organize long rest for infected pastures.
Foot-rot in sheep	Soft wet ground. Overstocked land. Buildup of bacteria in muddy gateways. Neglected and overgrown feet.	Sheep slightly lame. Sheep lying down more than usual. Slight infection of feet. Overgrown feet.	Sheep very lame. Sheep grazing on its knees. Hind feet drawn up towards front. Loss of condition, bad infection of feet. Sickly smell.	Pare feet regularly. Keep sheep moving onto clean pasture. Formalin foot-rot bath or treat with creosote. Slaked lime in muddy gateways.	Pare feet regularly. Treat with antibiotic aerosol and stand on concrete to dry. Cull really chronic sheep.
Orfe in sheep	Bacteria or virus infection.	Red, scabby lesions on top of hoof.	Lesions spread to lips and udders. Sheep become lame.	Antibiotic aerosol or vaccination.	Antibiotic aerosol or vaccination.
Mastitis in ewes	Too much milk soon after lambing. Ewes and lambs turned onto too rich grass too soon.	Ewe slightly lame behind. Lamb not allowed to suck. Hard quarter.	Ewe very lame and in poor condition. Quarter turns purple and begins to fall off. Pus.	As for cattle.	As for cattle. Sell ewe fat at weaning.

Table 7 cont.

Disease	Probable causes	Symptoms		Treatment	
		Early	Late	Early	Late
Twin lamb disease (pregnancy toxaemia)	Sudden change in nutrition close to lambing causing drop in blood sugar in ewe.	Weak, lack of appetite, unusual carriage, appears stupid.	Tremors of head and neck, partial blindness, refuses to rise.	Prevention of paramount importance; supplementary feed to ewes near lambing particularly in bad weather. Glucose injection intravenous or subcutaneous.	
Fly strike	Fly's eggs laid in wound or dirty or wet wool: maggots hatch out and eat into the skin or into rotten feet.	Tail twitching, attempt to get at infested spot with mouth, dark patch in wool.	Exaggerated twitching, wool comes out, open sore exposed, grazing stops.	Prevention by fly dipping 4 to 6 weeks after shearing and again in August.	Remove surrounding wool. Treat maggot with neat Dettol, dress with sheep dip or ointment.
POULTRY Red mite	Old buildings. Poor hygiene. Could be brought in with new birds.	'Salt and pepper' markings on perches at night. Human skin irritated after handling birds.	Birds become debilitated.	Regular cleaning and disinfecting of poultry houses, perches, nest boxes using creosote and paraffin mixture. Dusting birds with louse powder.	
Fleas	Old buildings. Poor hygiene Could be brought in with new birds.	Fleas picked up by humans.	Birds become debilitated.	Regular cleaning and disinfecting of poultry houses, perches, nest boxes using creosote and paraffin mixture. Dusting birds with louse powder.	

often the warning signs are missed or ignored, the condition becomes acute or even chronic, and the cure both costly and troublesome.

New-born animals are vulnerable to many infections, particularly on heavily stocked farms and in indoor conditions. Older beasts may be put under stress through high production or malnutrition (or a combination of both) and thus lose their natural resistance to disease.

The diseases of animals, their prevention and cure are described in detail by experts, who all agree that the building of health and prevention of disease are of first importance, and that removal of the cause is preferable to treatment of symptoms. When it comes to treatment, however, opinions tend to vary: at one extreme vaccinations and antibiotics are given routinely 'just in case' and, at the other, they are rejected out of hand. It seems to me, as a practical farmer, that there must be a balance; that both the old, traditional remedies and the more spectacular cures have their time and place. A cow showing the first signs of mastitis, for example, will respond readily to starvation, garlic, and cold water; an advanced case may require antibiotics. The older remedies demand time and patience, but tend to produce fewer complications; the new veterinary science is certainly quicker-acting and more convenient for large-scale application. I strongly recommend three books on animal health, each containing wisdom from a different viewpoint, each of which can serve the smallholder well: W. R. Wooldridge: *Farm Animals in Health and Disease*; F. Newman-Turner: *Herdsmanship*; and J. de B. Levy: *Herbal Handbook for Farm and Stable*.

Table 7 shows causes, symptoms, and suggested treatment of the most common diseases I have met. For other diseases one should consult the reference books mentioned earlier.

The foundation of animal health is sound nutrition and avoidance of stress. A modest yield and profit from animals that remain healthy is better than record yields that ruin their health.

The key to effective treatment of disease is keen observation, accurate diagnosis, and early action before the condition becomes acute. Any manifestation of disease should be a warning that our husbandry is slipping.

7 The smallholder's dairy

Many smallholders will want to keep a house cow and make butter, cheese, and yoghurt. The design and layout of the dairy will depend upon the scope of your dairying enterprise. If you run a small commercial herd of cows or goats and sell milk or yoghurt, you must obtain a licence and your premises must conform to government standards. For the same herd where only butter, cheese, or cream are sold – the skim milk and whey being used on the premises – no licence is required (apparently TB and brucellosis can be spread via untreated milk but not in cream, butter, or cheese): similar premises but different equipment are needed. When one or two cows or goats are kept purely to meet domestic needs, a much smaller set-up with very simple utensils will suffice.

Milk, especially when warm, is an almost perfect medium for the culture of any bacteria that gets into it. A diseased udder, dirty teats or hands, improperly sterilized milking machines, pails, churns, strainers, or dusty air can all infect milk and cause it to go off. The first principle of dairying is therefore strict attention to hygiene at every stage from cowshed to kitchen. Cooling, as soon as possible after milking, checks the multiplication of any bacteria and delays souring.

I keep two cows to provide milk, butter, cheese, and yoghurt for my family, and to rear calves. Surplus butter and cheese are sold at the farm gate. Although I need no licence, I have developed a dairy and routine that are very simple and yet completely hygienic.

Milking routine

My cowshed is of the old-fashioned byre design where the cows are chained by the neck and stand on concrete. In winter they are tied in at night on deep straw bedding. Before morning milking I muck out, swill down the floor, and clean off any muck and bedding from the cow's flanks, belly, and tail. I then wash my hands and arms and the cow's udder in hot water, massaging her thoroughly to trigger a let-down of the milk. I always milk with my sleeves rolled up to the elbow to avoid dust getting into the pail. The milk is then strained into a skimming pan, covered with a lid and wet towelling to cool it, and the cream skimmed off after twenty-four hours. The milk pail and strainer are then washed and rinsed in cold water, using a dairy brush. The outside of the bucket is scrubbed clean of milk and muck with a second brush. The cold wash is followed by a hot wash with a little detergent and a second cold rinse. The washing of all dairy utensils is done in a deep galvanized iron sink.

When the cows have been turned out for the day, the byre is again swilled down ready for evening milking which follows a similar routine.

Dairy layout

My dairy, which is quite separate from the cowshed, is on the shady side of the house. Adjoining the deep sink is a large draining board. The skimming pail and cream pail are kept on a formica-topped table backing onto a tiled wall. Under this table is a large shelf for milk pails and the cheese vat. Bolted to one end of the table is a steel, hand-operated rotary butter-churn that holds about $4\frac{1}{2}$ litres. Wooden shelves provide space for maturing cheeses. Next door to the dairy is a cloakroom where the dairyman can scrub himself up.

Equipment

The minimum requirements for a smallholding with one house cow are:

1 udder-washing bucket
2 udder cloths
1 milk pail (9-litre stainless steel is best)
1 strainer with filter papers
2 cream skimming pans with lids and towelling covers
1 cream skimmer
a dairy sink with hot and cold running water
2 dairy scrubbing brushes
storage pails or vats

Additional equipment may include:

single-unit milking machine
butter-churn
butter-worker
scotch hands
patting-up board
cheese vat
curd knife

cheesecloths
cheese press and weights
cheese moulds
cheese shelves
dairy thermometer
milk cooler
cream separator

Butter

Most of our cream is skimmed off to make butter. We use wide, shallow skimming pans. I prefer stainless steel (expensive) or enamel (hard to find) pans to aluminium, which react with lactic acid producing undesirable salts. After a year or two, aluminium pans (which are by no means cheap) are eaten right through and start to leak.

The cream is accumulated for up to a week (or until we have $4\frac{1}{2}$ litres), stirring each new lot thoroughly into the whole. Before making butter, all the equipment – churn, working board, scotch hands, table-top, and draining board – is scalded

with boiling water. The cream is tipped into the churn and the
temperature, which is critical, is adjusted (using iced or hot
water) to suit the ambient air temperature (see Table 8) and,
in so doing, the cream is thinned to a point where the wood of
the scotch hands can be seen through it.

Table 8. Churning temperatures

Air temperature		Churning temperature	
°C	°F	°C	°F
18	64	11	52
16	61	12	54
14	57	13	55
12	54	14	57
10	50	15	59

The cream is then churned until the fat globules separate from
the buttermilk and come together into crumbs that look like
scrambled egg. This will occur within twenty minutes if the
temperature is right. The buttermilk is drained off and the
butter washed through with five or six pails of cold water to
remove the last traces of buttermilk, ensuring that the butter
will keep. The butter is then tipped from the churn onto a
working board and worked by hand until all traces of water
have been removed. Salt may be added to taste during the
working, by sprinkling it onto the working board (approximately one teaspoon per 500 g.). The butter is then weighed
and patted up, using scotch hands, into 250 g. pats, decorated,
and wrapped in grease-proof paper. Stored in a refrigerator, it
will keep for six weeks. You need about 4½ litres of cream to
make about 2 kg. of butter.

Milking machines and cream separators are great labour-
savers for anyone milking five or more cows. They are
expensive and take a great deal of washing up. Milking by
machine requires a very high standard of husbandry, other-

wise cows' udders may be damaged and mastitis spread through a herd.

Channel Island cattle are poor converters of carotene; thus their cream and butter is always of a rich yellow colour. Butter from some other breeds may look better if colouring is added.

Cheese

We make a farmhouse cheese which is something between a Caerphilly and a Cheddar, using half skim and half whole milk from not more than four milkings. As with butter, all equipment is scalded and the cheese-maker scrubs up his hands and arms very thoroughly. The milk (up to 23 litres) is heated in a vat up to 32°C. Standard cheese rennet, 1 teaspoonful to 23 litres, mixed in eight times its volume of warm water is then added and well stirred in. The vat is then covered with a towel or blanket away from draughts and left for approximately 40 minutes for the curd to form. A drop of cold water is shaken onto the curd; if it rests on top like a pearl, the curd has set and is ready to cut, if not, it must be allowed more time.

The curd is cut with a long thin knife in two directions to release the whey. After cutting, it is stirred gently with the hand until it is evenly broken up into pieces the size of sugar lumps and left to stand for ten minutes for the whey to flow; the whey should then be a yellowish-green clear liquid. The whey is then gently ladled off into a pail through a cheesecloth, the curd getting firmer and drier as ladling proceeds. The faster this can be done the better. The curd is then tipped into the cheesecloth and hung for one hour to drip. The curd, now solid, is weighed, roughly cut, and salted by hand (one teaspoon per 500 g.). The cheesecloth is then spread into a cheese mould and filled with curd, which is pressed down gently and evenly into the corners. A single layer of the cheesecloth is folded over the top of the curd, a metal or wooden disc 'follower' placed on top and the mould placed in the press with a 7–9 kg. weight on top.

Cheeses are turned daily in the press, put in dry cheesecloths and under increasing weights of up to 25 kg. for five days. They are then removed from their cloths and placed on clean wooden shelves in the dairy where they are turned and wiped off daily for a fortnight. Thereafter they are left to mature for up to five months. They may be eaten 'green' after six weeks, but develop better flavour and texture after longer maturing.

Cottage cheese is made by pouring naturally soured milk through a cheesecloth and allowing it to drip for three days, scraping it 'outside-to-middle' each day. All soft cheeses have a very short shelf life.

Lactic or yoghurt cheese is made by pouring yoghurt through a cheesecloth the same as for cottage cheese. This cheese is excellent for making cheesecakes.

Whey or Norwegian cheese is made by reducing the whey until it becomes a sticky, slightly crystalline solid. It is pressed into small bowls. It contains the milk sugars, protein, and a little fat. Its sweet taste is unlike other cheeses, and it may take time to acquire a liking for it.

In making these cheeses, we have settled on a simple system and on types suitable to our own needs. We use no starters, no bandages. The texture and flavour of our cheeses are governed by the condition of the grass (a good curd can only be achieved in the growing season from May to September), the breed of our cows, and the particular bacteria endemic in our dairy and equipment. A vast range of options and refinements are open to the cheese-maker, but, unless he sets up an expensive establishment with a controlled environment, his end product will be limited by the three factors I have mentioned.

Yoghurt

Yoghurt is made by bringing whole milk or skim slowly up to the boil to kill off the indigenous bacteria, cooling it to about 46–49°C and adding a selected bacterial culture (normally Lactobacillus Bulgarius or Streptococcus Thermophilus) in their place. The result is a strong curd.

The original yoghurt is started with a liquid or dried culture according to makers' instructions. Subsequent batches are innoculated with yoghurt from the previous day's lot.

The culture is smeared around the sides of a china bowl at room temperature (about two tablespoonsful of yoghurt to two litres) and left exposed to the air for a few minutes. The prepared milk (46–49°C) is then added, stirring thoroughly. The bowl is then covered with a dinner plate and warm towels, and left for three hours in a warm, draught-free room. It is then transferred to the refrigerator.

Yoghurt may not be sold except under licence.

Skim milk and whey

Surplus skim milk and whey are excellent sources of protein for poultry and pigs. I store this surplus outside in an old milk churn where it sours quickly, and feed it as required.

The maxim in the dairy should be:

- Keep it clean
- Keep it cool
- Stick to a strict routine.

8 Vegetables, fruit, and flowers

The growing of vegetables, fruit, and flowers, on a smallholding may be a full-time commercial venture or merely a small side-issue for domestic consumption. Modern, specialized farmers have tended more and more to rely on the milk-roundsman and the supermarkets, finding any aspect of self-sufficiency a time-wasting obstacle to profit-making. But genuine profits relate to value for money and the quality of living they can yield. The mixed smallholder is in a position to feed his family on fresh, top-grade produce, which, unless he is very foolish, must be better value for money (or value for his time) than goods that have passed through the hands of wholesaler and retailer.

Most smallholders will attach great importance to their gardens and orchards, and will, in spite of needs and overriding priorities elsewhere, feed their families mainly on home-grown fruit and vegetables.

The smaller the holding, the greater the emphasis must be on horticulture (intensively farmed small animals such as rabbits, mink, and poultry always excepted), and the easier it should be for the owner to concentrate his time and energy on gardening.

Gardens and orchards vary enormously in potential depending on altitude, aspect, rainfall, soil, fertility, and growing season. The books I refer to in the Suggested Reading section contain excellent advice on the principles, but should never be read too literally in matters of detail such as planting dates. Each gardener must build up his own wisdom from experience, and adapt the principles to his own conditions.

Rotations and fertility

The first principle of good gardening is to build up and maintain the fertility of the living soil so that the garden can go on and on producing an abundance of healthy vegetables. Eating garden vegetables is, in itself, extractive husbandry as human wastes, unlike those of farm animals, are seldom returned to the soil. Unless you put back the equivalent of what you take out, fertility will drop, and with it crop yields. Some plants are 'hungry' for nitrogen (potatoes), others can fix it from the atmosphere and return it to the soil (legumes). Some need calcium (brassicas) and others do best on 'last year's' fertility (carrots). Some species are 'good companions', others are incompatible.

Most domestic vegetables are annuals or biennials (the latter are consumed in their first season – if left they set seed in their second year), and do not thrive when sown again where they grew last year.

It follows, therefore, that some sort of crop rotation, tied to fertility-building, must be devised to produce the best crops year after year.

The principles of fertility, including compost-making, have already been discussed in Chapter 4. The basic requirements in the garden are the same as in the field:

● Good drainage
● A more or less neutral pH
● A plentiful supply of humus
● The prevention of mechanical damage to the soil structure, especially in wet weather.

In a relatively small garden, it is not as difficult to achieve these conditions as it is in the field. I have found that some crops are prone to garden pests but grow well in my roots field. Garden potatoes are attacked by eelworm, whose egg cysts are apparently made more fertile by passing through the gut of an earthworm. By building up fertility and the earthworm count in our beds, we seem to be building a problem! In spite of companion planting with onions and garlic, our garden carrots are attacked by fly, and garden cauliflower (*inter alia!*) are damaged by our prolific slugs.

Our farm, Brynoyre (should be Brynoer, meaning Cold Hill) is more than 600 feet above sea level, has a high rainfall and a late, short growing season. In spite of our efforts to drain it, the soil becomes waterlogged in winter and the high proportion of fine silt makes the surface liable to cap after heavy rain. The soil is deep, acid, and potentially fertile, the structure improving year by year. To counter these conditions we have gradually evolved the system I shall now describe. In it you may discern the basic principles and adapt them to your own conditions.

We divided our ground, about one-tenth of an acre, into four equal plots which we double dug over a period of three years. Plot 1 was for early potatoes, Plot 2 legumes, Plot 3 brassicas, and Plot 4 root vegetables, to follow each other in that order. Compost was applied before potatoes and legumes, calcified seaweed (lime) before brassicas, and the roots had to get by with 'last year's fertility'. Patches of perennial Welsh and tree onions, rhubarb, artichokes, and strawberries intruded into the rotation. We found that by walking between the rows to harvest or hoe, we packed the soil very hard and impeded aeration and drainage. In 1978 we adopted the raised bed system: 5-foot beds with 15-inch paths between.

In 1981 we abandoned early potatoes (we had tried polythene tunnels and glass cloches) and made over Plot 1 to marrows. We now grow maincrop potatoes, peas, and carrots on the ridge in the roots field.

Succession crops

To get any vegetables started before June in our conditions, we had to build a greenhouse where seeds could be sown in

March and April, to be pricked out into boxes, hardened off in a cold frame, and planted out in the beds in May. We tend to sow and prick out many more brassicas than will fit into Plot 3, but plant them out very close to be transferred to the roots field headlands after hoeing is completed in July. By this means we have grown some excellent late cabbages, sprouts, and cauliflower.

We generally find it difficult to clear the sprouting broccoli, which is at its best in April and May, in time to get early root crops sown.

In theory we always plan a succession of sowings of peas, French beans, carrots, and spinach as well as Winter Density lettuce and Chinese cabbage which cannot go in before midsummer. But haymaking, shearing, hoeing, and corn harvest are the enemies of our garden hopes, and we generally end up with more weeds than we like to see. Nevertheless, our garden provides practically everything we need, and we sell or give away a good deal as well.

Greenhouse

Our greenhouse, 9 ft. by 20 ft., is a lean-to built into a south-facing 'L' corner of our house. Existing door and window heights prevented us from setting the south face and roof at the optimum angle. The house wall at the east end blocks the early morning sun. After growing very leggy seedlings for two years, we lined the inside (stone) walls with panels covered with aluminium foil, and painted everything else white. The increase in light was remarkable, and the greenhouse did not become quite so hot in the early afternoon. Instead of trestles and platforms, the main growing areas are concrete-floored 'wet-beds', with between one and two inches of sharp gravel in the bottom and seed compost, growing compost, or loam above. All watering is done via perforated hoses in the gravel, and comes up to the plant roots by capillary action. This helps to avoid surface capping, although the very top occasionally dries out on a hot day. Above the wet beds are shelves for seedling boxes. We use a small paraffin heater to keep the temperature above freezing during March and April. In one corner of the greenhouse is a drying cupboard, through which warm air from an electric fan heater is blown. We use it for drying our own seeds.

Indoor tomatoes and early lettuces are grown in the wet beds, as well as brassicas, leeks, runner beans, marrows, courgettes, cucumbers, and garden flowers for later bedding out. The indoor tomatoes (we have given up the uneven struggle with them outdoors) are grown against the house wall, with French marigolds between them to discourage the white fly.

Cold frames

We harden off seedlings in a crude, plastic-covered cold frame before bedding out. We occasionally get a frost in early June that could kill the marrows and runner beans.

Orchard

Our quarter-acre orchard consists mainly of apple trees: earlies and keepers, eaters and cookers, planted as two-year-old semi-standards in 1977. Into each hole we put some bone meal, hoof and horn, and wool daggles from our shearing shed, covering them with upturned turf before planting the trees. In 1978 and 1981 we removed the turf out to the drip-line (that is, out to the extremities of the overhanging branches) and have mulched this area with part-rotted compost every autumn. The trees have been carefully pruned each spring, and are now fruiting well. In 1980 we started breeding earthworms, and in 1981 mulched all the apple trees with compost from the worm pits, full of egg capsules and young worms. This compost was covered with grass mown from between the rows of trees. We hope that by increasing the earthworm population, the soil structure, drainage, and aeration will be improved (the orchard tends to lie very wet in winter).

Between the rows of apple trees, we are gradually establishing beds of blackcurrant bushes, and elsewhere about the garden are victoria and greengage plums, cherries, pears, and quinces.

Birds take our cherries and soft fruit before they are fully ripe, so we have to pick them as they ripen.

Pests

I am inclined to believe that vegetables and fruit grown on healthy, living soil acquire a measure of immunity to disease and pests. We certainly get pests – slugs are the worst – but the trees and plants seem to grow through them and yield very satisfactory crops. Occasionally we are forced to use pyrethrum or derris for green or white fly, but for the persistent pesticides we feel no need at all.

Storage

Space should be reserved in the granary or similar dry room for vegetables and fruit. Some will be stored in racks, some in sand or peat. In all cases provision must be made to protect them from frost. My fruit and vegetables are stored on the granary floor above the cattle byre which seldom drops below freezing. In very cold weather I cover the racks with a blanket or straw.

9 Machinery

The Problem

The *genuine* profitability of a smallholding depends upon your ability to keep your costs below your sales, and to produce enough livestock and crops to reap sufficient gain from this differential. The amount a smallholding can produce is limited by the land potential, and, within this potential, by your skill and energy. Assuming you are skilled, your time and energy are often the factors that prevent you from achieving your full land potential, and you are constantly looking for power sources and techniques that will extend your physical strength and efficiency to this end.

Your problem is to find the power and artifice at a price that truly increases your long-term profit.

Fixed costs, such as rent (or interest on land-capital investment) and rates, are beyond your control and probably do not vary much with the size of the farm. Investment in livestock and land improvement, on the other hand, may vary in proportion to acreage and have a direct effect on profits. Machinery and equipment cannot be direct profit-makers within the farm, yet may enable a single-handed farmer to achieve more profit-making work. Whether or not an extra profit *is realized* will depend on the 'rent' (that is, interest on capital), and depreciation, maintenance, and running costs of the equipment you buy. And herein lies the small-holder's dilemma: unless you over-capitalize on machinery, you may not have the energy to work your land to its full potential.

So you have ten acres and you buy a tractor, plough, and

other implements; enough equipment, perhaps, to work fifty or even a hundred acres. But for most of the time your tackle is idle and depreciating, and is seldom worked to capacity. Because you own a tractor, you are always tempted to buy more labour-saving implements to go on it, and before long you find that your limited land potential cannot support your investment.

It is extremely difficult to get the balance right and keep it so.

The options

To avoid over-capitalization on machinery you have several courses open. You can hire contractors, borrow from neighbours, grass-let your land to other men's livestock, buy and repair cheap second-hand equipment, adopt human or animal power and third-world techniques, or seek small, appropriate power sources and implements. You may try to minimize depreciation by doing your own modifications and repairs, and by meticulous routine maintenance.

On the other hand, you may deliberately over-capitalize on sophisticated machinery and become a part-time contractor working for others. Your problem here could be keeping a balance between your holding and outside work.

Contractors

Because their equipment is more fully utilized, and they generally have better maintenance and repair facilities, contractors can carry out most operations more cheaply than farmers. Their profit, however, uses up the differential, so there is not much to be gained in the long run, from relying entirely on them. In the short term, however, there are definite advantages. You can avoid the capital outlay during your first year's operations while you assess your holding and decide what equipment you need. You can save your own time without the administrative trouble of employing farm workers. If a contractor's machine breaks down, it can probably be replaced or repaired quickly.

There are disadvantages as well. Contractors must be paid without undue delay. Cultivations and harvesting tend to come in a rush when the weather is right, and it can be frustrating for a smallholder, probably at the end of the queue, when his work is not done promptly. Most contractors are geared to large farms, employ heavy tractors and tackle and work at high speed – not always the most suitable combination for a smallholder who wants to plough shallow, or harvest tiny fields.

There are some operations such as lime- and slag-spreading, rotovating, baling, and combining, where the cost of equipment is disproportionately high in relation to a smallholder's annual needs, and here the contractor is the obvious choice.

Neighbours

Most beginners receive ready help from good neighbours until such time as they can reciprocate. There is no better way to learn about your land than by leaning on your neighbours. One can soon find out what machines and implements are needed and most suitable. But friendships wear thin on one-sided borrowing.

There is a lot to be said for sharing the more expensive implements; but, from my own experience, I consider it unwise for two farmers to own jointly any one piece of equipment. If each item is owned by one person, he is responsible for maintenance and has first call on it; and there is less chance of friction. In this way it is possible to make fuller use of machinery without such high capital investment.

Stick and dog

It is possible to farm with practically no machinery at all. By confining your enterprise to grass and livestock, you can eliminate all cultivations, arable crops, and harvesting. A holding may be 'grass-let' for the summer with cattle and the winter with sheep. The maintenance of fences, gates, drainage, and water supply, and keeping an eye on the animals are your only responsibilities. If you chose, you could rest your land between grass-lets or shut it up completely during the winter and get away for a holiday. The alternative to grass-letting is to manage your own 'flying' flock or herd, matching the animals to the grass available. In this case it might be necessary to make some hay, when it would be sensible to employ a contractor to cut and bale, and hire or borrrow other hay-making equipment from neighbours.

This system requires good grassland management, and it is always wise to lay down strict conditions and stocking rates before signing a contract to let. Except on good fattening pastures, the profit to be made from grazing animals is not great, but the outlay and risks are minimal.

Old equipment

It is still possible to pick up second-hand bargains at sales and, over a period of years, to equip a smallholding with all the tools required for mixed farming. But the second-hand means worn, and not many items in working order are sold below their value. The bargain-hunter must know what he is looking for and be a good judge of mechanical soundness; he must also be able and equipped to modify and repair his bargains to make them workable.

I have equipped my holding with very old implements over a period of many years and at extremely low cost. To do so I have gone to a great many farm sales and more often than not have left empty-handed, as I do not bid above a real bargain price for anything. I have a very well-equipped workshop and forge where I can make parts and do repairs, but even so I find it difficult to keep my museum pieces running year after year.

And, as the years go by, potentially workable bargains are harder to find.

I would not recommend this method of equipping a holding unless you have a delight in renovating machinery.

Animal power

The use of animal power can be viable for a smallholder who knows what he is doing, particularly if he has a growing-up family who shares his enthusiasm. But working with horses, mules, or bullocks is time-consuming.

Harness and animal-drawn equipment are expensive and hard to find, but make progressively more economic sense as the cost of heavy modern machinery and fuel soars.

Horse-drawn implements were developed to a very high standard in the late nineteenth and early twentieth centuries, and were operated by competent horsemen who spent many years learning their skills and accepted the excessively long hours involved. Today, no such tackle is made in Britain, although it can still be acquired, at a price, in a few East European countries. Some very simple, light, implements are being made mainly for bullock draught in developing countries, which could be useful with ponies or cobs on smallholdings. These are mainly cultivation and sowing tools. Reaping tackle – mowers, turners, and binders – require heavier horses and cannot easily be justified on very small farms, but they are being imported from Poland. If two or three neighbouring smallholders each owned a cob, they might well acquire a complete set of equipment between them.

If you kept work-horses you would probably have to buy in their hay and oats, otherwise you would have to reduce profitable livestock by an equivalent amount.

Appropriate technology

For a great many years the farms of Britain have been growing larger and fewer. The implements to work them have become heavier, more complex, and costlier. Tractor and implement manufacturers, finding that their greatest profits come from more powerful machines, have progressively abandoned the simpler, smaller equipment most appropriate for small farms. The smallholder cannot possibly afford modern, heavy equipment, and is faced with the choice, already mentioned, of doing without, employing contractors, or making do with old, worn machines; that is, unless he can find a source of modern equipment specially designed for small farms.

Power sources

The power to operate farm implements is provided by tractors, self-propelled garden-type cultivators, animals, or man himself. Animal-draught tools can be drawn very inefficiently by tractors, but otherwise each of the above power sources needs its own implements. You should therefore decide which source to adopt, and stick to it.

A new range of mini farm tractors, mainly Japanese, is now becoming available with complete sets of hydraulic-lift cultivating and harvest tools, link boxes, and trailers. These are considerably smaller than the old TVO Ferguson 20 hp machines, which were the ideal size for the small farmer. These new tractors and their equipment are expensive, and have not yet established the nationwide service and spares organization of the older firms.

Self-propelled, pedestrian-operated cultivators are now well proven on horticultural holdings where tillage and limited transport are the predominant tasks. They are not suitable for hay-making, muck-spreading, and the heavier farm operations. A commercial-sized cultivator costs nearly as much as a mini-tractor.

It is perfectly possible to run a smallholding on animal power supplemented by hand tools if you have the skills and time. A well-trained draught-horse is cheaper to buy than a mini-tractor and has a longer working life. The cost of his food over the year compares favourably with tractor fuel and repairs for similar work done. The problem today is to find a complete set of implements to match one size of horse.

If you rely entirely on manpower you will be very limited in the scope of your farming, although there are now many excellent labour-saving hand tools for the vegetable garden. The cost of hand tools is negligible compared to engine- or animal-drawn equipment: for example, a seed-fiddle can be used to broadcast cereals, roots, and grass almost as quickly as they can be drilled, yet the cost is only about 5 per cent of the cheapest corn drill. There are many jobs on a smallholding for which hand tools are the obvious choice, whatever main power source is adopted.

Tools

Since 1960 a great deal of research has gone into the design and production of tools and equipment for small-scale husbandry. The late Dr E. F. Schumacher's philosophy of 'Small is Beautiful' is embodied in the Intermediate Technology Development Group which has catalogued and assessed a vast range of simple, inexpensive equipment, mostly suitable for the developing world, although they could become more appropriate in Britain as the energy crisis bites. You would be well advised to study the ITDG catalogue in order to put together a balanced set of equipment.

The smallholder in commercial horticulture will probably need to be fully mechanized with tractor, self-propelled cultivator system, or animal-drawn equipment, whereas the mixed smallholder with a relatively small domestic vegetable garden and orchard will almost certainly use hand tools.

The amount of fixed and domestic equipment will depend on the degree of self-sufficiency to which you aspire. If you are producing primarily for sale, very little will be processed on the farm; if you are farming for your own subsistence you will need to thresh, winnow, and grind your corn, make butter and cheese, chop roots, cut chaff, roll oats and barley, all of which require special equipment.

Shelter for equipment

Farm machinery, tools, and equipment are extremely expensive to buy. It goes without saying that they must be properly maintained if they are to last. If exposed to the elements in an overgrown rick-yard, corrosion is rapid and routine maintenance more likely to be neglected. An agricultural mechanic once said to me '75 per cent of the repairs I carry out could have been avoided by protection from the weather and the regular use of a grease gun.' This appalling waste should not occur on any farm; no smallholding can possibly afford it.

Implements and tractors

Implements take up a great deal of floor space and do not stack neatly. Most of them require little head room but nothing can be stacked on top of them. It is therefore a waste of space to store them in tall buildings such as Dutch barns. Long, low, open-fronted sheds, back to the prevailing wind, make the most convenient shelters; but even in these it is difficult to arrange storage so that you can extract the implement you want quickly. Some rarely used equipment such as corn and root drills may be raised by chain block and suspended from the barn rafters; others, such as balers, may be sheeted under strong tarpaulins outside.

Tractors may be easier to start and are certainly easier to service if kept under cover. Ideally the tractor shed, mechanical workshop, and oil and grease stores should be planned as a unit with electricity laid on.

Tools

Farm and garden hand tools, veterinary medicines and

equipment, and other odds and ends need a room or shed where they can be stored neatly in their proper places. Without it, tools are left lying around and cannot be found when needed. Under the pressure of even normal farming operations, it is too easy to leave tools and equipment lying around unless you have the proper place to store them and the self-discipline to use it. In a crisis you may lose a beast through not knowing where to find veterinary gear.

Work places

At best a farmer's lot is long hours and hard work. You are nearly always short of time. Any arrangements that make your routine tasks less laborious and more efficiently done must merit consideration. You may delegate some jobs to outside contractors (cultivations, machinery repairs, hedging and so on) or employ workers to do them within the farm. But this involves extra expense, and you will always be looking for ways of doing all the work more efficiently yourself.

Farm workshop

A well-laid-out and equipped workshop allows routine maintenance and repairs to be carried out quickly on the farm at minimum expense. If you are an inventive type and good with your hands, you will be able to equip yourself to do all but the most complex jobs, including modifications and the development of new ideas.

The first requirement is a mechanical bay with bench, vice and tools for normal adjustments to tractor and implements. Electric welding facilities are a valuable addition. This bay is best sited next to the tractor shed and should have concrete floor space to accommodate one or more implements being repaired.

A carpenter's shop is next in importance for repairs to house and buildings. Simple brick-laying, masonry, and concreting tools are often required as buildings and facilities are modified to meet new requirements; a tractor- or hand-driven cement mixer is a great boon. Electrical tools including a battery bench and charger take up little space, and are often needed.

On my own farm I have all these facilities as well as harness-making tools and a complete smithy. Space is also required to store nails, bolts, staples, spare parts for machinery, paint, and a host of other items used in the workshop. Good lighting is very important, as is a power source for drills, grinders, and similar machines.

10 Making ends meet

There came a great spider
Who sat down beside her . . .

If Miss Muffet was prepared to be frightened away (as many of my full-time students were when their eyes were opened to the reality of farm work) the spider might be seen as a benevolent and useful creature who saved her from real disaster later on. But Robert the Bruce was made of sterner stuff, and I like to think that the spider preferred his role in that story.

In this book the reader will now recognize three themes. The first is 'HUSBANDRY' or what, in my opinion, a farmer ought to do. The second is (pseudo) 'ECONOMICS' or what most farmers are forced to do. The third I call 'PIG-IN-THE-MIDDLE', or my efforts (by no means a financial success) to reconcile the first two. I believe that most commercial farmers are far from happy with the current state of affairs, and long for the day when *genuine* good husbandry will pay. Perhaps we all expect too much for our long hours and cannot see why others, who do much less, should earn much more. So we compromise on what we know in our hearts to be right and somehow make ends meet. This may appear to work on a medium-to-large farm; but for the average smallholder it is becoming very difficult to subsist, much less make a profit. This was not always so, and I ask myself why can I not afford today what was common practice a few years ago?

In the 1950s I stayed for some weeks *en famille* on a thriving small Tyrolean hill farm. They had no car, television, radio, or telephone and were largely self-sufficient in their economy.

They worked closely with their neighbours whenever extra labour was needed, and formed a very close-knit family and village community. Nowhere have I ever met healthier or happier folk.

This sort of life should be possible today, but how? This presents a very deep problem, obviously beyond the scope of this book. All I can do is to give a few indications as to how you may manage to make ends meet on today's smallholding. What steps can we take to close the financial gap *without betraying the basic principles of good husbandry*?

Improvements to existing farm economy

Within the economy of a very small farm, your options are limited. By reducing costs, improving efficiency, working harder, keeping accurate accounts and records to show you where you go wrong and where opportunity beckons, increasing output (both in quantity and quality), growing new products, and by better marketing, you can certainly reduce your losses and you might even make a small profit, but ultimately your land potential is the deciding factor. You will almost certainly need other sources of profit to make a fair living, and the more lucrative these are, the less will be the pressure on your farming.

Outside employment

A part- or full-time job away from the holding may be undertaken by one or other partner. On a very small holding this might be possible without reducing output below land potential; but in most cases it would be necessary to simplify the enterprises and reduce the intensity and work-load within the farm. I know one part-time medical consultant and two schoolmasters who also farm successfully.

Subsidiary enterprises

It is possible to run a subsidiary enterprise within the farm. This might be achieved with little disturbance to the farming routine and scope, or could involve a complete reorganization. It is obviously very difficult if there is only one of you, but quite possible if one member of the partnership or family takes the responsibility for it.

If the farmhouse is suitably equipped with spare bedrooms and bathrooms, bed-and-breakfast guests can be very profitable without a great capital outlay or undue extra work. My wife provides bed-and-breakfast only – never midday or evening meals – and our guests seldom come into the sitting room or violate our privacy. The home-produced breakfast is very cheap and greatly appreciated. Carefully placed advertisements in tourist centres and publications ensure a steady flow of trouble-free visitors. We have never had a 'B and B' sign at the farm gate.

A neighbour of mine has completely reorganized his farm as a pony-trekking centre which has paid much better than mixed farming. This enterprise needs special skills, a considerable capital outlay on stables and chalet accommodation, and puts a severe limit on conventional farm operations during the tourist season. Out-of-season the ponies are non-productive and do the farm no good.

Another neighbour runs a similar project providing horse-drawn caravan holidays. A third, unwilling to face the treadmill of unending expansion, has sold his dairy herd and converted his farm buildings into flats for summer visitors. The land is now used for grass-letting and hay for sale; and a fourth runs a caravan site. Three friends of mine in West Wales have built up a thriving market for organically grown fruit and vegetables. Their husbandry, packaging, advertising, and marketing are highly professional. These are imaginative schemes that have worked: not all do.

For more than a decade my own solution to the problem was to turn my farm into a training centre running courses in many aspects of small-scale farming. To achieve this my wife and I formed a charitable trust which enabled us to raise funds to set up and run the project. To begin with we trained young Tibetan refugees from India, and later accepted students from Britain. Our ten-acre farm, together with a neighbour's 45-acre holding, was equipped and worked up to demonstrate almost every aspect of self-reliant smallholding and its associated crafts. In spite of a strong demand for the training we offered and a measure of initial success, we failed to gain official recognition; our students were not eligible for public authority grants and had to be subsidized out of project funds. We thus failed to make the project pay and had to give up early in 1982. Before the project finally closed we made plans to open the farm to the public with demonstrations, farm trails, picnic sites, tea room, information centre, and bookshop. There was every indication that the scheme would succeed, but we were unable to get planning permission because our access roads were too restricted.

Identifying our true needs

So far we have considered schemes for increasing income. Perhaps now is the moment to ask ourselves why we need so much? What is essential and what luxury? How far could we balance the books by doing without, and what would that imply? There is a tendency for yesterday's luxury to become today's necessity. During my childhood in Western Canada, we lived for some years without electricity, running water, central heating, telephone, or motor car. The wireless was in its infancy and TV unheard of. Yet we were well fed and clothed and enjoyed full, civilized, and happy lives. During the War we survived deprivation and hardship without loss of health or morale. We managed this because *we were all in the same boat.* Surely we could manage even better in peacetime if only we could find the real incentive.

To do without the affluence of modern society today implies a *voluntary* self-denial. I may be prepared to accept a 'lower' standard, but is it morally right to expect my family to do likewise?

There will always be a few outstanding husbandmen who succeed against all the odds, and there are small parcels of exceptional land that will yield a good living (mainly from horticulture) to a competent man who is prepared to work extremely hard. But today these are the exceptions. For the majority smallholding cannot be considered a viable alternative in the present economic conditions. But the future of our affluent society is in some doubt and who can tell when it will make political sense to encourage large numbers of young people to return to the soil?

If we cannot adjust our values and our demands the time may not be far off when we have no option. In either case the smallholder will eventually come into his own and face the great challenge of becoming self-reliant on the land. The sooner the better.

Notes

1. Elkington, W. M., *et al.*, *The Smallholder's Handbook* (The Bazaar, Exchange and Mart, 1918).

2. Evans, T. W., *Land Potential* (Faber, 1951). My italics.

3. McCarrison, Sir Robert, *Nutrition and Health* (Faber, 1962), quoted in S. B.-I. Sweeny, 'Health from the Living Soil' (*International Journal of Environmental Studies*, 1977).

4. Stewart, V. I., 'Soil Structure', *Soil Association Quarterly Review*, **1**(3), 1975.

5. Stewart, V. I., 'Soil Damage and Soil Moisture', I. H. Rorison and R. Hunt (eds), *Amenity Grassland – an Ecological Perspective* (Wiley, 1980).

6. Balfour, E. B., *The Living Soil and the Haughley Experiment* (Faber, 1975), p. 108.

7. Edwards, C. A. and Lofty, J. R., *Biology of Earthworms* (Chapman and Hall, 1977), pp. 21 *et seq.*

8. Stapledon, R. G. and Hawley, J. A., *Grassland* (Oxford University Press, 1927), pp. 126–7.

9. Watson, J. A. S. and More, J. A., *Agriculture: The Science and Practice of British Farming* (Oliver and Boyd, 1956), p. 611.

10. *The Importance of Colostrum*, Livestock Farming Supplement on Calf Rearing, August 1972.

Suggested reading

General

Elkington, W. M., *et al.*, *The Smallholder's Handbook*, The Bazaar, Exchange and Mart, 1918.

Elliot, R. H., *The Clifton Park System of Farming*, Faber and Faber, 1944.

Henderson, G., *The Farming Ladder*, Faber and Faber, 1956.

King, F. H., *Farmers of Forty Centuries*, Rodale Press.

McCarrison, R., and Sinclair, H. M., *Nutrition and Health*, Faber and Faber, 1961.

Newman-Turner, F., *Fertility Farming*, Faber and Faber, 1951.

Rainsford-Hannay, F., *Dry Stone Walling*, Stewartry of Kirkudbright Dry Stone Dyking Committee, 1976.

Sweeny, S. B.-I., 'Health from the Living Soil', *International Journal of Environmental Studies*, 1977.

Sweeny, S. B.-I., *Self-sufficient Smallholding*, Soil Association, 1976.

Sweeny, S. B.-I., *Working up a Smallholding*, Blackthorn Press, 1980.

Watson, J. A. S., *The Farming Year*, Longmans Green, 1938.

Watson, J. A. S., and More, J. A., *Agriculture: The Science and Practice of British Farming*, Oliver and Boyd, 1956.

Soil and fertility

Balfour, E. B., *The Living Soil and the Haughley Experiment*, Faber and Faber, 1975.

Barrett, T. J., *Harnessing the Earthworm*, Faber and Faber, 1949.

Bruce, M. E., *Commonsense Compost Making*, Faber and Faber, 1973.

Darwin, C., *Humus and the Earthworm*, Faber and Faber, 1966.

Edwards, C. A., and Lofty, J. R., *Biology of Earthworms*, Chapman and Hall, 1977.

Evans, T. W., *Land Potential*, Faber and Faber, 1951.

Howard, A., *An Agricultural Testament*, Oxford University Press, 1943.

204   *Suggested reading*

Minnich, J., *The Earthworm Book,* Rodale Press, 1977.
Temple, J., *Worm Compost,* Soil Association, 1979.

Crops

General

Lockhart, J. A. R., and Wiseman, A. J. L., *Introduction to Crop Husbandry,*
  Pergamon Press, 1978.
Ministry of Agriculture, *Tractor Ploughing,* HMSO, 1971.
Sweeny, S. B.-I., *Smallholder's Harvest,* Soil Association, 1977.

Grass

Voisin, A., *Better Grassland Sward,* Crosby Lockwood, 1960.
Watson, S. J., and Smith, A. M., *Silage,* Crosby Lockwood, 1956.

Vegetables and fruit

Fisher, R., and Yanda, W., *The Food and Heat Producing Solar Greenhouse,*
  John Muir Publications, 1976.
Hills, L. D., *Grow your own Fruit and Vegetables,* Faber and Faber, 1974.
McCullagh, J. C., *The Solar Greenhouse Book,* Rodale Press, 1978.
Royal Horticultural Society, *The Vegetable Garden Displayed,* RHS, 1972.
Royal Horticultural Society, *The Fruit Garden Displayed,* RHS, 1974.
Shewell-Cooper, W.E., *The Complete Greenhouse Gardener,* Grenada/
  Mayflower Books, 1979.

Livestock

General

Hagedoorn, A. L., *Animal Breeding,* Crosby Lockwood, 1948.
Levy, J. de B., *Herbal Handbook for Farm and Stable,* Faber and Faber, 1975.
Sweeny, S. B.-I., *Looking at Livestock,* Soil Association, 1977.
Thear, K. (ed.), *The Complete Book of Raising Livestock and Poultry,* Martin
  Dunitz, 1981.
Wooldridge, W. R., *Farm Animals in Health and Disease,* Faber and Faber,
  1960.

Cattle

Livestock Farming Supplement on Calf Rearing, *The Importance of Colostrum*, August 1972.

Newman-Turner, F., *Herdsmanship*, Faber and Faber, 1952.

Russell, K., *The Herdsman's Book*, Farming Press, 1956.

Sheep

Bowen, G., *Wool Away*, Whitcombe and Tombs, 1949.

British Sheep Breeders' Association, *British Sheep*, BSBA, 1968.

Thomas, J. F. H., *Sheep Farming Today*, Faber and Faber, 1966.

Horses

Ewart-Evans, G., *The Horse in the Furrow*, Faber and Faber, 1975.

National Federation of Young Farmers' Clubs, *Farm Horses*, NFYFC, 1936.

Poultry

Balfour, E. B., *Keeping Poultry on a Small Scale*, Soil Association Technical Booklet No. 2, Soil Association, 1980.

Eley, G., *Home Poultry Keeping*, E. P. Publishing, 1976.

Malcolm, B., *Successful Poultry Farming*, John Bellows, 1930.

Thear, K. (ed.), *The Complete Book of Raising Livestock and Poultry*, Martin Dunitz, 1981.

Worthington, J., *Natural Poultry Keeping*, Crosby Lockwood Staples, 1974.

Pigs

Thear, K. (ed.), *The Complete Book of Raising Livestock and Poultry*, Martin Dunitz, 1981.

Goats

Hetherington, L. V., *Home Goat Keeping*, E. P. Publishing, 1977.

Bees

Mace, H., *The Complete Handbook of Beekeeping*, Ward Lock, 1976.

Dairying

Ministry of Agriculture, *Farmhouse Butter Making*, Advisory Leaflet 437, HMSO, 1957.

Ministry of Agriculture, *Farmhouse Cheese Making*, HMSO.

Ministry of Agriculture, *Hygiene Requirements for the Sale of Milk*, HMSO.

Tools and equipment

Intermediate Technology Development Group, *Tools for Progress*, ITDG.

Useful organizations

ADAS (Agricultural Development and Advisory Service), Ministry of Agriculture, Whitehall and Local Government offices
Provides information and advice on all aspects of agriculture. HMSO publishes leaflets and booklets on almost every aspect of farming.

Agricultural Training Board: National Agricultural Centre, Stoneleigh, Kenilworth, Warwickshire: also regional offices
Runs local practical courses in many farming skills.

Alternative Technology Information Group: Flat 10, 85 Westbourne Terrace, London, W2
Collects, correlates, and disseminates information pertaining to every aspect of alternative technology.

John Arden: Stinhall Farm, Stiniel, Chagford, Devon
Runs courses in managing and driving harness horses.

Centre for Alternative Technology: Llwyngwern Quarry, Machnylleth, Powys
Carries out research into and demonstrations of various aspects of alternative technology, including ambient energy sources (wind, water, solar, and methane power), methods of insulation, organic food-growing, recycling of resources, local manufacture, and self-reliance. Open to the public. Restaurant, bookshop, and information centre.

Chase Organic Seeds: Benhall, Saxmunham, Suffolk
Produces a mail-order catalogue of organic vegetable and flower seeds; also many other gardening supplies.

The Country Trust: Denham Hill Farm, Quainton, Aylesbury, Bucks

Provides an informed introduction to the countryside, conferences, lectures, exhibitions, and practical courses in making bread, butter, and cheese, and in the growing and use of vegetables.

County College: Well Hall, Alford, Lincs

Runs one-year residential and correspondence courses in horticulture and agriculture.

The Good Gardeners' Association and **The Horticultural Training College**: Arkley Manor, Arkley, Nr Barnet, Herts

Runs: one-year courses on the growing of all types of food crops, for students intending to work abroad as missionaries; one-year courses on growing fruit, flowers, and vegetables (including under glass) for students who plan to work in Britain; two-year diploma courses in horticulture.

Henry Doubleday Research Association: Convent Lane, Bocking, Braintree, Essex

Carries out research into improved methods of organic horticulture, aimed particularly at amateur gardeners and smallholders. There is a quarterly newsletter to members with results of experiments on their own experimental and demonstration plots, and a mail-order service of books, organic fertilizers, pesticides, and other materials.

Intermediate Technology Development Group: 9 King Street, Covent Garden, London, WC2

Founded by the late Dr E. F. Schumacher (author of Small is Beautiful).

Aims to: stimulate new thinking and action on the part of both rich and poor countries so that overseas development finance can work more effectively for the benefit of the whole of the community to which it is directed; supply basic and applied research in Britain and training facilities on the spot to provide the type of industry best calculated to relieve unemployment and poverty in the developing countries. ITDG publishes Tools for Progress, *a catalogue of tools and implements, many of which are suitable for smallholders.*

The McCarrison Society: 23 Stanley Court, Worcester Road, Sutton, Surrey

Founded to further the work of the late Sir Robert McCarrison.

Aims to: spread the knowledge of the relationship between nutrition and health through regular meetings, conferences, and liaison with other scientific societies; encourage and initiate research projects in the field of health and nutrition, and to encourage the teaching of nutrition within medical schools; collect and publicize available evidence relating to aspects of nutrition and health.

Organic Growers Association: Aeron Park, Llangeitho, Dyfed

Provides advice and technical information on organic horticulture from national and international sources. Researches into every aspect of organic growing as well as the development of the marketing of organic produce, including presentation and packaging.

Practical Self-Sufficiency and Small-Scale Supplies: New Leys Farm, Widdington, Saffron Walden, Essex

Produces a monthly magazine, Practical Self-Sufficiency. *Sells supplies, tools, and equipment for the smallholder, and has an extensive bookshop.*

Project Equipment Ltd: Industrial Estate, Rednal Airfield, West Fenton, Oswestry, Salop

Designs and manufactures multi-purpose animal and light tractor-drawn implements at very reasonable prices. Implements are mainly for the developing world, but have significant application to small farms anywhere.

Self-sufficiency and Smallholding Supplies: The Old Palace, Priory Road, Wells, Somerset

Catalogue and shop for supplies, tools, and equipment for the smallholder, including an extensive bookshop.

The Small Farmers' Association: Cilpyll, Bwlchllan, Dyfed

Aims to: stabilize or increase the existing total number of farms; make them more accessible to new entrants; further the interests of existing smaller farmers; co-operate, where appropriate, with all people and organizations interested in the affairs of the countryside.

The Soil Association: Walnut Tree Manor, Haughley, Stowmarket, Suffolk

Aims to: bring together those concerned with the vital relationships between the health of soil, plant, animal, and man; offer guidance on the principles of organic husbandry; sponsor the marketing of produce meeting specified standards; gather and disseminate information on all aspects of its concerns. The SA has an excellent bookshop.

The Yarner Trust: Dartington, Totnes, South Devon

Runs long and short residential courses providing practical training in small-scale mixed organic farming and the self-sufficient economy.

Glossary of farming terms

abomasum: the fourth, or true digestive stomach of a cow

ADAS: Agricultural Development and Advisory Service provided by the Ministry of Agriculture

agistment (tack): the sending away of livestock (commonly lambs for the winter) to another farm where conditions are more favourable

aspect: the direction of the slope of a piece of land; e.g. 'south-facing'

brimming: in season, on heat, when referring to a sow

butt: the butt of a grass sward is the leafy part near the ground in which most of the nutrition is found

bye-tack: *see* in-bye

byre: a cowshed with stalls where cattle are tied by the neck, with mangers in front of them and a dunging passage behind

cash crop: a crop grown specifically for sale

catch crop: an additional or 'bonus' crop grown between the harvesting of one and the sowing of another main crop in a rotation

cattle mesh (pig wire): galvanized iron wire-mesh fencing with eight horizontal strands and with vertical wires at 6-, 9-, or 12-inch spacing

cladding: timber, sheet-metal, or asbestos weather-boarding on the outside of a farm building

clamp: a storage-pile of roots (turnips, potatoes, etc.) covered with straw, bracken, or earth for protection against the elements

cob: a stocky riding/driving horse

coedcae: *see* in-bye

concentrates: concentrated feed consisting of grain or mixtures of grain, pulses, etc. without the bulky straw. This feed may be of pure home-grown cereals or complex mixtures from exotic sources. It may or may not contain added minerals, fish-meal, growth additives, and anti-biotics.

crush: a narrow stall (which may be portable) in which a cow may be completely restrained

dagging: removing daggles with shears

daggle: a hard knob of wool and dung on the hindquarters of a sheep

dam: mother

dredge: a mixture of cereals or cereals and pulses grown together for animal food

extractive crop: a crop removed from the land (for sale or feeding elsewhere) where the land suffers a loss of fertility as a result (as opposed to a grazing crop)

fallow: bare ground (ploughed or cultivated) on which no crop is growing

ffrith: *see* in-bye

flocculation: the process whereby very small silt or clay particles are collected into more or less coherent aggregates or crumbs

foggage: grass which has been allowed to grow up and fall over without being cut or grazed

folding: the controlled grazing (generally of sheep) over a crop of roots by use of hurdles or electric fence

FYM: farmyard manure consisting of animal dung, urine, and bedding

gilt: a young female pig

grower: a young pullet that has not reached point of lay, or a cockerel before being fattened

guano: phosphate fertilizer derived from the droppings of seabirds

humus: a product of decomposition of plant and animal residues through the agency of micro-organisms. The chemical composition of humus is determined by the nature of the residues from which it is formed, by the conditions of its decomposition, and the extent to which it is decomposed

in-bye: rough, enclosed land adjacent to open, common land (also coedcae, bye-tack, ffrith)

kibbled: very coarsely ground grain

leaching: the washing out of nutrients from the soil by water

legume (pulse): a member of the pea or bean family, the seeds of which are formed in pods and the roots of which include nodules which can fix nitrogen from the atmosphere

ley: a grass sward that has been sown to a prescribed mixture

lime: a loose term covering the calcium required in the soil to attain a certain pH. It may be applied as ground limestone (calcium carbonate), slaked lime (calcium hydroxide), quicklime (calcium oxide) or calcified seaweed

lodge: when a cereal crop grows too tall and weak-stemmed, it may fall over (particularly in a storm) and is said to 'lodge'

loessic: loessic silts are accumulations of silt particles deposited by the wind

NPK: nitrogen, phosphorus, and potassium; considered by many to be the principal essential plant nutrients

nurse cow: a cow used to suckle one or more calves which are not her own

nurse crop: a cereal or root crop with which are sown grasses and clovers that are thus established and protected in their early growth

oil cake: a concentrated feed made up of one or more types of oil-seeds such as ground nuts, sesame, rape, mustard

organic: organic husbandry embodies the concept of organic wholeness applied to the vital relationships between soil, plant, animal, and man with special reference to health (in every aspect of that relationship). It envisages health as a positive process which is profoundly influenced by the interdependence of all life, primarily through the medium of nutrition

panning: the consolidation of a layer of soil rendering it impervious to moisture. A surface 'pan' is caused by the sole of a plough sliding repeatedly over the subsoil at the same depth

pH: a measure of the acidity or alkalinity of the soil (literally a measure of the concentration of the hydrogen ion): pH4 is extremely acid, pH7 is neutral, pH9 is very alkaline

poaching: the indentation of the soil surface by animals' hooves, tractor wheels, and implements, particularly in wet conditions

pulse: *see* legume

race: *see* shedding race

root crop: turnips, swedes, mangolds, sugar beet, potatoes, carrots. It is sometimes applied loosely to other brassicas which are not strictly root crops, such as kale, rape, and cabbages

rumen: the first stomach of a ruminant animal

scour: diarrhoea

shedding gate: a gate at the end of a shedding race by which sheep may be sorted into two separate pens

shedding race: a narrow passage up which sheep may pass single file: in it they may be counted, dosed, or injected

shoddy: waste wool scraps from a woollen mill

slag: an artificial source of phosphorus (P_2O_5) formerly a by-product of the steel industry (basic slag)

soil structure: the manner in which the various particles of soil (sand, silt, clay, and organic matter) are knit together into a crumb structure or tilth

soil texture: the ingredients of a soil: sand, silt, clay, and organic matter

stook: a stack of four or more sheaves of corn piled together in a field to dry and harden

store: a growing beast or sheep which is kept on a relatively low plane of nutrition before being fattened or 'finished'

swathe (windrow): a line of grass or corn after mowing with scythe or mower, normally 4–5 ft. in width. When tossed up for drying by the wind it is referred to as a windrow

tanalize: to impregnate posts or stakes with tannins to preserve them against rotting

tedding: shaking up hay with a fork or tedder-machine to let in air and thus speed drying

tiller: when several stems of corn shoot up from one root, it is said to 'tiller'. This tends to happen when the young plants are grazed off before being allowed to grow up to head

trace element: one of a great number of mineral elements present in the soil in extremely small amounts. They have vitally important effects on the health of plants and animals living in or on the soil

tumbledown: a grassy sward that has reverted from an arable crop or ley to indigenous species of grasses and weeds

tup: a ram

tupping: the mating of sheep

undersown: when grass is sown together with an arable crop (cereals or roots), it is said to be 'undersown'. The arable crop may be referred to as the 'nurse crop'

weaner: a young pig after weaning

windlass: a system of ropes (or chains) and pulleys designed to put a tension on a wire or cable

windrow: *see* swathe

zero grazing: a forage crop that is cut and carried green to be fed to stock

Index